Workshop Machinery

Workshop Machinery

Alex Weiss

Special Interest Model Books

Special Interest Model Books Ltd.
PO Box 327
Poole, Dorset, BH15 2RG
England

www.specialinterestmodelbooks.co.uk

First published 2010

© 2010 Alex Weiss

ISBN 978 185486 260 0

Printed and bound in Malta by Progress Press Co. Ltd

Contents

Acknowledgements

It is impossible to write a book like this without the help of a great number of people. I am indebted to many companies who agreed to the use of their photographs to help to illustrate the text. The list includes all of the following people: David Cummings of All Machine Tools, Hugh of Amadeal Ltd, Jacqueline Hunt of Axminster Tool Centre, Daniel De La O of Bolton Hardware/Tools, Howard Barrett of Boxford Ltd, Martin Brown of BriMarc/Proxxon, the staff of Chester Machine Tools, the staff of Chronos Engineering Supplies, Colin Childs of Cowells Small Machine Tools Ltd, Nicola J. Denford of Denford Limited, Christa Vergnes of Emco Maier GmbH, Alan Simmonds of Arc Euro Trade, Bill Griffin of Grifftek Technical Services, Melinda Sweet of Grizzly Industrial Inc, Julian Davis of HME Technology Ltd, Steve and Chris Holder of Home and Workshop Machinery, Tanja Hug of JET/Walter Meier Manufacturing, Richard J. Kinch, Robert Bertrand of Lathemaster Metalworking Tools, Tony Griffiths who runs the web site www.lathes.co.uk, Dr Kurt Daley of MicroProto Systems, Lester Caine of Model Engineers Digital Workshop, Ingrid Weiskopf of Optimum Maschinen GmbH, Peter Morrison of Peatol Machine Tools, Bryan Tate of Pro Machine Tools Ltd, Chris of RDG Tools Ltd, Allison of Rondean Machinery, Craig Libuse of Sherline Products Inc, Robert Thompson of Smith Hamilton Group, Matt Washkow of South Bend Lathe Company, Craig Daley of Taig Tools, Greg Jackson of Tormach LLC and Roger Warren of Warco. If there is anyone I have forgotten to thank, then I apologise for the error, which is entirely mine.

In addition, my grateful thanks go out to my model-engineering soul mate, Kevin Walton, who has provided endless support to me through out the writing of this book. Last but not least, my dear wife has sustained me throughout the time it has taken me to put together the contents of this volume.

Every effort has been made to ensure that all of the technical information in this book, as far as is humanly possible, is accurate. It is also hoped that almost all current machines have been included.

Introduction

For anyone contemplating setting up a new workshop from scratch or about to acquire a major new item of equipment for an existing workshop, a number of decisions must be made before going ahead with any machinery purchases. Unfortunately, it is easier to choose the right new piece of equipment or accessory once some experience has been gained in its use. This is a classic 'Catch 22' situation. Hopefully, this book will enable both newcomers and experienced model engineers considering acquiring a lathe or milling machine to make the right choice to suit their personal needs. Regrettably, there is no best buy since this inevitably depends on the actual needs of the user.

What will be made?

The first issue to consider is the type of items that are likely to be made in the workshop. There is the world of difference between the requirements of a model aero-engine constructor, someone intent on building a 7¼" steam locomotive, anyone who makes or repairs machine tools, those who restore full-size traction engines and at the other extreme people who build or restore clocks, watches or delicate scientific instruments. So it is very important to think about the requirements of current modelling or full-size engineering interests

Figure 1. *A scale Gipsy aero engine requires precise engineering.*

Figure 2. *A 7¼" locomotive is a large model to produce and transport.*

Figure 3. *Working on a full-size traction engine requires large machinery.*

as well as any possible future changes of interest. The size of what is being built is a key consideration in the purchase of most machine tools. Think about the likely maximum diameter and greatest length of components to be machined. To illustrate

Figure 4. *Clocks and watches need finely finished, delicate parts made to high accuracy.*

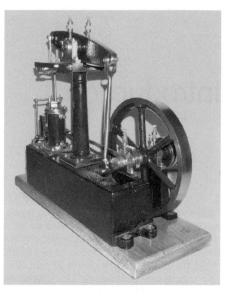

Figure 5. *A well-built example of the Stuart beam engine.*

this point, the lathe reputed to have the world's largest distance between centres is at the Rosyth shipyard in Scotland and can accommodate propeller shafts up to around forty metres (130 feet) long. On lathes used for turning turbo-generator casings, the operator can sit within the lathe to change tools. So it is essential to think not just about current ambitions but also possible changes in interests in the future. As well as ensuring any lathe or milling machine is of adequate size, it is also true that making delicate work on a large machine can be very frustrating.

Some model engineers are happy to make working models where the ability to run is the primary aim. Others are very competitive and may aim for a museum-quality standard. The first can often be completed with relatively rudimentary equipment; the latter may cause the user to show more concern for the quality of

Figure 6. *Cherry Hill's exquisite model of a Blackburn 1863 engine.*

Figure 7. *While few people will try to make their own lathe or milling machine, tool cutters and grinders are popular home-build projects that are similarly complex.*

the machine tool. The complexity of any planned model or any full-size item of equipment is also an important factor, as it affects the machinery that will be needed to complete its construction.

How much use will equipment get

Some thought needs to be given to the amount of use that any machine will have to undertake. Irregular evening use is less demanding than forty or more hours use per week that might be undertaken by a dedicated pensioner. The total usage will vary depending on the actual piece of equipment; a combination lathe/mill is likely to get more use than an individual single-function machine.

There is a mass of equipment that has been designed for the model engineering market. There is also much more that is primarily aimed at the mass-production industrial market. While this is perfectly satisfactory for home workshop use, it tends to be a relatively over-engineered and over-priced solution for home needs.

Financial implications

Equipment budgeting can be a difficult subject, especially if funds are limited. No-one is likely either to wish or be able to afford to buy all the machinery and accessories they need at one go. It may take years or even decades to acquire all the items desired. Even then, changes in technology will result in improved, more-capable or even novel items becoming available at regular intervals. Choosing priorities and then planning an initial and an annual budget for workshop equipment should enable items to be purchased in a sensible order.

When looking at the prices of machine tools, it is important to realise that the inclusion of a good range of accessories can significantly increase the price of a bare lathe or milling machine. In addition, payment terms can vary considerably and may change with alterations in market conditions; becoming more favourable in recessionary times.

Make or buy

Many model engineers enjoy making workshop equipment. *Model Engineers' Workshop* is a magazine dedicated to the construction of workshop equipment. However, home building a complete lathe or milling machine from scratch is a rare occurrence. *Making Small Workshop Tools* WPS 14 and *Useful Workshop Tools*

Figure 8. *An elderly Myford ML7, once the perfect choice for the home workshop. Excellent second-hand examples can be found and spares are still available for most models. Photo courtesy Home and Workshop Machinery.*

WPS 31, both by Stan Bray as well as Harold Hall's *Model Engineers' Workshop Projects WPS 39* all provide details of how to make a number of useful devices some of which are handy adjuncts to lathes and milling machines. Home-made items are often significantly cheaper than purchased equivalents but require considerable amounts of time to be invested in their construction. The choice of whether to make or buy any specific item will depend on the facilities and skills of builders as well as their particular interests, time available and financial status.

New or second-hand

Major savings can be made by buying second-hand equipment. However, the old adage "caveat emptor" or buyer beware comes into play. Without a considerable amount of experience, or a friend with such knowledge, it can be very difficult to decide whether a second-hand item like a lathe is suitable for purpose and is a good buy. On the other hand, the saving can be quite substantial but it is important to know the current price of an equivalent new item to avoid paying more than the

Figure 9. *This old Colchester Student lathe may have seen better days but is probably capable of restoration to near as-new condition. Photo courtesy Home and Workshop Machinery.*

going rate. For some people, the challenge of restoring an old machine to tip-top condition is sufficient reason to justify its purchase. Many of the best brands of machine will provide sterling service and last a lifetime, even if bought second-hand. On the other hand, inexpensive new machinery, largely made in the Far East, may not be as long lasting, and some people have had experience of parts manufactured from a quality of steel that is inadequate for purpose.

Factory reconditioned units
There is a difference between a second-hand machine tool and one that has been refurbished by the original equipment manufacturer and is sold with a warranty. All of the worn parts should have been renewed giving an "as new" performance.

Equipment quality

There is a big difference between buying the best and obtaining value for money. The quality of workshop machines vary considerably, with those manufactured by

4

Figure 10. *The Chester DB11 GVS lathe is an imported machine of Chinese manufacture.*

Figure 11. *Limited space in many small workshops is a key consideration that may make a multi-function machine, rather than a separate lathe and milling machine, a perfect solution. Photo courtesy Warco.*

the industrial nations generally offering greater precision and longevity than ones produced in the Far East. However, many companies in the industrial world are turning to China and other Asian countries for their manufacture. Furthermore, the importers of Far-Eastern equipment are driving up the quality and will get machines ready for use prior to delivery. Despite this, there are still occasions when these machine tools are assembled with, for example, poor quality nuts and bolts, though these are readily replaced.

An issue that is important both on new and second-hand machines is the amount of backlash in the operating controls. And anyone who has tried to turn the handle to move the cross-slide on a quality lathe and then a cheaper import should have immediately been able to feel the marked difference in quality.

Space implications

Of course, the space available in a home workshop will impact on the individual size and number of different machine tools that can be accommodated and, in some

cases, access to get new machines into the workshop may also be a factor. A few people, lacking a workshop, may opt for portable equipment which can be stored in a cupboard when not being used.

Forrest and Jennings in their book *Workshop Construction, WPS 23,* explain how to build a new workshop. But for the majority, a garden shed or part or all of a garage is likely to impose significant space restrictions. This means that for major items lack of space may restrict the choice of machine or, at the worst, mean that it cannot be housed without a workshop extension. This is often a reason for choosing a smaller machine than is ideal, or for selecting a multi-function item such as a lathe complete with a milling head.

It is also worth considering the environmental conditions in the workshop. Much equipment can be ruined by rust. This can be difficult if not impossible to prevent in some situations. Thus it may, in these circumstances, be best either to choose equipment that is less likely to rust or less expensive items that can be replaced if serious rusting does occur.

Figure 12. *Moving a one-ton Bridgeport mill requires a large and sturdy trailer as well as adequate lifting facilities. Photo courtesy Richard J Kinch.*

Transporting and lifting

It may not be easy to get a heavy item of equipment back home from the supplier and lift it into position in the workshop. Access to a suitable trailer and fork-lift truck or hoist may be essential. However, all other things being equal, a heavier and more rigid machine tool will be less prone to vibration and chatter, resulting in better quality parts being produced.

Access
A limiting factor on the size of equipment may be imposed by the width of the door into the workshop or the need to move equipment along a passage that has a ninety-degree turn or steps.

Personal recommendations

There is nothing quite like a suggestion from a fellow club member or another acquaintance who already owns the item being sought and whose advice can be

Figure 13. *A visit to a friend's workshop to examine their equipment and, if possible, use it is an excellent idea.*

trusted. Good advice can often provide a useful insight into its usability, reliability, quality and accuracy as well as whether it represents value for money.

Magazines like *Model Engineer*, *Model Engineers' Workshop* and *Engineering in Miniature* all provide equipment reviews from time to time. These are well worth reading or re-reading if they are relevant to a planned purchase. Similar information can often be found on the Internet.

It will, of course, be sensible to assess the skill and experience of anyone who makes a recommendation before acting on their advice; preferably not advice from the club "know all"!

Try before buy

It is clearly a good idea to get 'hands-on' any major piece of equipment before buying it. There are several ways in which this may be achieved. A visit to a friend's or fellow club-member's workshop is ideal if they already own the item that is planned to be purchased. This may well allow the opportunity for a little personal practice

Figure 15. *Visitors can examine both new and pre-owned machines at Myford's open days.*

Figure 14. *Have a good look around suppliers' stands at model engineering exhibitions and try the movement of the various handles.*

with the item of equipment. Alternatively, a visit to the manufacturer or to a local supplier may be a practical proposition. However, look and touch will normally be the practical 'hands-on' limit. Finally, there is the opportunity to see a wide range of potential workshop items for sale at model engineering exhibitions though again there will be a limit to how much can be tried.

Metric or imperial

An early decision that needs to be made is whether to go metric or imperial. This decision may already have been made with existing equipment either to one or the other standard. If starting from scratch, there are two considerations. The first is any personal preference for metric or imperial units. The second factor is that, on the one hand, the vast majority of existing model-engineering designs will use the imperial system, while on the other hand, imperial tools and materials are slowly becoming both increasingly difficult to find and expensive to purchase, apart from second-hand items. Also, designs of new models increasingly use metric units.

Outside the United States where imperial units are commonplace, it is a metric world, particularly in Europe although many in the UK still think in imperial units and, even in the US, the majority of large engineering companies have long ago embraced the metric system.

Throughout this volume, both metric units and their imperial alternatives are given (or imperial units with their metric equivalents); often figures quoted by manufacturers or suppliers. These conversions will rarely be exact but rather to the nearest round figure.

Equipment support

It is always worth checking what sort of warranty and return policy a manufacturer offers before making a large purchase.

A major supplier that is relatively local is often the best bet when it comes to spares and support for expensive items of new workshop equipment. However it is always worth trying to establish the degree of support available as well as its timeliness and cost.

In the case of second-hand equipment, it is useful to establish whether spares are still available and whether a support group exists. Club members and the Internet are useful sources of information. It is also noteworthy that spares prices for foreign-made equipment will reflect fluctuations in currency-exchange rates. These have been relatively volatile in recent years.

Single- or three-phase power

The vast majority of model engineers will want machines that will operate from a single-phase electrical supply (via a 13amp plug in the UK and its equivalent overseas). A few home workshops may have a built-in three-phase supply. For those without such a supply but who wish to run one of the larger machines that needs a three-phase electrical supply, one solution is to use a power converter. A number of specialist suppliers manufacture a range of suitable units and ones to suit almost all sizes of machine are offered.

Personal bias

Everyone is biased and it is important to recognise this fact. A few typical personal biases include:

1. A strong preference not to purchase foreign-built equipment.
2. Support for any supplier whose equipment has given good reliable performance in the past.
3. An inclination to believe the words in glossy magazine advertisements.
4. Accepting pressure from a sales person in a shop or at a model show.

Concluding what to buy

The topics so far have indicated the wide range of factors that should be considered before making any purchase decision. The choice is likely to be hardest for the most expensive items of equipment. There is no magic solution. It seems that some compromise is always needed. It is hoped that this book will provide a useful overview of what is available and which factors need careful consideration.

About this book

The purpose of this book is to help both beginners and more-experienced model engineers to choose the best machine tool to meet their particular and often very personal requirements. It covers most of the machines suitable for model engineers that are made in the UK, Europe and the United States as well as those imported from the Far East.

In those parts of the text dealing with American-built machines and equipment imported into the United States, both American terminology and spelling have been used whenever possible.

Unsurprisingly, the book starts by looking at a wide range of conventional lathes. It then provides a similar assessment of combination lathe/mills and also milling attachments that can be fitted to lathes. Chapter 3 provides details about the many machines that specialise in vertical or horizontal or both types of milling. Chapter 4 considers the growing number of computer-numerical-controlled (CNC) lathes and milling machines that are increasingly popular with modellers. The last chapter examines the accessories, such as cutting tools and holding devices, which either come with the machines or can be separately purchased.

The Conclusions provide details of how to look after machine tools. Following a list of useful names, addresses and web sites, there is a comprehensive index.

Chapter 1

Lathes

A conventional, as opposed to a Computer Numerically Controlled (CNC) lathe, is normally one of the first machine tools purchased. A classic lathe with appropriate tooling will turn, face, bore, drill, mill, fly cut, ream, polish, cut tapers, knurl and the majority will also produce metric and imperial threads. A few watch- and clock-maker's lathes may lack a lead screw and thread-cutting facility.

Choosing the right lathe is not an easy task as there are many alternatives and it is not made any easier since most model engineers only buy a lathe once in their lifetime; when they start to equip a workshop (referred to as a shop in North America). It is assumed the many factors indicated in the introduction have been reviewed and a decision made about the size, budget and quality of lathe desired, whether it will be built to metric or imperial standards, space in and access to the workshop, any preference for a new or second-hand machine and whether equipment support is an important factor.

The classic model engineer's lathe in the UK is a Myford but with the upsurge in Far-East products, the choice is wide open. There is little question that Asian lathes offer great value for money but they also require careful setting up. Often the design

and production quality are not as good as lathes made in Europe or North America. In those two regions, there are many indigenous machine-tool manufacturers and their products will often be preferred to lathes made in Eastern Asia.

Perhaps the first two critical issues are the centre height (swing in North America) and the distance between centres. The former defines the maximum diameter that can be turned and the latter the maximum length. In both cases, the first thought must be the type of work that will be undertaken on the lathe. What is the largest diameter and what is the longest length of any metal that will have to be turned; not necessarily at the same time? A gap bed, with or without a removable section, can be an advantage as it increases the diameter of a component, like a wheel, that can be turned. In addition, the spindle bore will affect the maximum diameter of a long item that can be held in the chuck.

The type of lathe bed does not in itself matter. Lathes with flat beds, V-beds or round beds, the last either single or twin, should all be satisfactory but the width of the bed is a factor; generally the wider the better. The headstock bearings may be ball or taper-roller bearings (easy to replace) or plain bearings found on older lathes and

Figure 1. *The Wabeco D2000 E lathe has a centre height of 110mm (4¹/₃"), a distance between centres of 300mm (11¾") and speeds from 30 to 2,300 rpm. Photo courtesy Pro Machine Tools Ltd.*

on modern high-precision ones. Often a mixture of bearing types is employed.

A cross slide, ideally with T-slots and with a separate bolted-on top slide that can also be angled, adds flexibility. With a suitable angle block the top slide can also be used as a vertical slide.

At this point, review the amount of space that is available in the workshop or other area where the lathe will live and whether it will be bench-mounted, free-standing or small enough to be readily portable. As a

general rule, a lathe that is slightly too big is a far better proposition than one that is just too small. A screw-cutting ability is important if threads are to be cut. A half nut fitted on the lead screw will allow rapid repositioning of the cross slide. The vast majority of lathes have this facility.

Consideration needs to be given to the list of accessories available for the lathe such as 3- and 4-jaw chucks, a face plate and the way they are fitted to and removed from the headstock. Chuck back plates

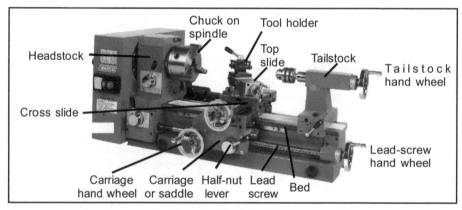

Figure 2. *The main parts of a typical thread-cutting lathe. Photo courtesy Warco.*

that screw onto the headstock nose are quicker to change than those that have to be unbolted from the back plate and the new chuck bolted back on. However, screw fittings may, with time, over tighten and be difficult to remove. Equally, they can come loose when using back gears. Camlock chucks are by far the easiest to change. A vertical slide and four-way tool post are also popular accessories. Asian lathes tend to come with a good range of accessories included in the price while lathes made in the industrialised nations tend to offer these as optional extras at additional cost.

The accuracy of the lathe and items like chucks should be checked, though modern machine tools are likely to meet the needs of most home users. In addition, any component made in the home workshop only has to fit other home-made components, unlike industry where items made on opposite sides of the globe have to fit accurately together. For the average home use repeatability figures of 0.025mm (0.001") should be acceptable.

For any new or second-hand lathe, the feel of the hand wheels when moving the carriage and cross slide provides some useful information, but also depends on gib-strip adjustment that should be easy to carry out. The types of handle and the ease of reading and zeroing their associated dials are important considerations.

Acquiring a second-hand lathe requires a few specialist skills in order to allow an accurate assessment of the quality of the offering to be made. It is essential to be able to check for wear in the headstock, the bed and slides as well as the accurate location of the tailstock. Are the Morse tapers in the head and tailstocks in good condition and what about the drive motor, belts and gears? Is there a full set of change wheels for turning different threads? Are spares still available for the lathe?

Figure 3. *Many older Myford Super 7 lathes, made in the UK, are still in daily use by the model-engineering fraternity.*

A single-phase motor supplied with the lathe is readily operated from the mains electricity but a three-phase motor will require an inverter unless a suitable supply already exists. In the case of a single-phase motor, it can come as a fixed-speed unit (the lathe speed is altered by changing the pulleys on which the drive belts run or through varying the ratios in a gearbox) or with an electronic speed controller to drive a direct-current (DC) motor. Of course, the speed of rotation of any three-phase motor is readily varied.

The choice will also depend on any preference for electronic speed control; sometimes not ideal at the lowest speeds but easy to vary compared with belt changing when a different speed is needed. The lathe gearbox of a 'geared-head lathe' adds to the expense while allowing for rapid speed changes. A lathe with back-gear much reduces spindle speed and increases torque; desirable when turning large-diameter components and when screw-cutting. A tumble reverse is used to alter the direction of travel of the carriage under power. In neutral it allows rotation of the headstock spindle without driving the lead screw.

Mention must also be made of small lathes, typically employed by clock and watch-makers where the headstock is

Figure 4. *The working parts of a modern Boxford lathe showing a quick-change tool post and speed controls.*

turned by a bow. There are even a few foot-treadle-operated lathes in existence that are occasionally found second hand.

So what is the desirable speed range for the lathe itself? First, all other things being equal, a large lathe needs to run more slowly than a small one. So for a large lathe, a speed range of 25 to 2,000 rpm may be fine; for the smallest ones 100 to 5,000rpm or more is desirable.

The major suppliers of lathes depend on the user's location. In Europe, Germany has long had the reputation for building quality machine tools while in countries all over the world, imported East Asian (usually Chinese) lathes will provide a great deal of machine for a comparatively low cost. In the United States, relatively large lathes ranging from 10" to 22" swing (125mm to 280mm centre height) are popular for a nation where the population density and the house construction favour relatively large shops. Most lathes, including those sold in the US, can be purchased either in metric or imperial versions, mostly with the ability to cut both types of threads.

Finally, as a potential buyer, try and list the essential and desirable parameters of any lathe before making a purchase.

New European and US lathes

There are numerous lathe manufacturers around the world. Examples include Myford in the United Kingdom, South Bend in the United States, Wabeco in Germany and Emco Maier in Austria. The machines made in the industrialised nations are shown in alphabetical order of manufacturers. Companies like Axminster, Chester and Warco in the UK and Harbor Freight and LatheMaster in the United States all import lathes of Far Eastern manufacture.

While UK and European companies tend to quote centre height and distance between centres in metric units, American practice is to refer to a lathe by a pair of numbers, for example 9 x 30, both imperial units, meaning a swing over the bed of 9" (twice the center height) and a distance between centers of 30".

Boxford

This UK company is deeply involved in CNC but also manufactures three relatively large manual lathes, the **280**, **330** and **330TR**. The first has a centre height of 140mm (5½") and the other two 170mm (6²/₃") The distance between centres of the **280** and **330** can be either 500mm, 750mm or 1,000mm (19²/₃", 29½" or 39¹/₃") while the **330TR** only comes with the larger two centre distances. The lathes are 1,285, 1,535 or 1,785mm (50²/₃", 60½" or 70¼") long, depending on the centre distance and are 667mm (26¼") wide. Weights vary from 330kg (727lb) to 450kg (992lb), depending on the model and the bed length selected. The spindle nose fits 3-DI Camlock chucks. The lathes are driven by 1.1kW (1½hp) or 1.5kW (2hp) three-phase motors and fitted with gearboxes providing half, plus 50%, double and 8:1 multipliers. There is a range of different accessories, some of which are pre-fitted to the **330TR** lathe.

Figure 5. *The Ceriani David lathe can be supplied with longitudinal power feed. Photo courtesy Chester.*

Figure 6. *The Cowells 90CW clock- and watch-maker's lathe is only 500mm (20" long). Photo courtesy Cowells Small Machine Tools Ltd.*

Ceriani

This Italian company makes a lathe that comes in three variants; the **David 201, 202** and **203**. The **201** uses manual feed while the **202** comes with two automatic feeds and the **203** with four. The gap-bed lathe has a centre height of 100mm (4") and a centre distance of 500mm (19²/₃"). Powered by a 560W (³/₄hp) motor, the spindle is belt driven to give a speed range of 100 - 1,800rpm. The machine measures 100 x 500 x 350mm (4" x 19²/₃" x 13¾") and weighs 80kg (176lb). The lathe comes with a 100mm (4") 3-jaw chuck, a tool-holder turret with two tool holders and a set of spanners. The lathe is available across Europe and is imported into the UK by Chester Machine Tools.

Cowells

Cowells manufacture a range of three small precision lathes; the **90E Basic**, the **90CW** and the **90ME**. All three are made in England, and each has a gap bed. They all have centre heights of 44mm (1¾"), gap-bed swings of 120mm (4²/₃") and a distance between centres of 203mm (8"). The **90E Basic** and **90ME** use a belt drive and pulleys for six direct and six indirect spindle speeds from 60 - 2,100rpm. The **90ME** has a 90W motor while the **90CW** has an electronic control giving speeds from 40 - 4,000rpm from a 125W (¹/₆hp) motor. The motors are mains powered to

European or US standards. The **90E Basic** weights 8.5kg (19lb) and is supplied without a motor. The other two lathes weigh 22kg (49lb) and have a T-slotted cross slide and taper-turning top slide.

Accurate to within 0.005mm (0.0002"), the **90CW** is an easily portable clock-, watch- and instrument-makers' lathe that accepts 8mm (¹/₃") horological collets in the headstock and tailstock spindles. The **90ME** has back gearing and is optimised for screw cutting. Both have an aluminium base with integral speed control.

The **90CW** has a sensitive lever-feed tailstock, 60-division indexing headstock spindle, an 8mm (¹/₃") collet housing in the headstock and tailstock, a quick-change, height-adjustable tool post, a tailstock with an offset capability and a carriage lever feed (or lead screw). There is an almost endless list of accessories.

Emco

This Austrian company make several lathes increasing in size from the **Unimat 4** that has a centre height of 46mm (1¾") and a distance between centres of 200mm (8") with a 95 watt motor giving eight speeds from 130 - 4,000rpm. It weighs only 6kg (13¼lb). The **Unimat Basic** has a centre height of 54mm (2") and the distance between is centres of 194mm (7²/₃"). An 80W (¹/₈hp) motor gives a speed range of 20 - 2,200rpm. It

13

Figure 7. *The Emco Compact 5 lathe is popular for teaching basic lathe use. Photo courtesy Home and Workshop Machinery.*

weighs just 13kg (24lb). The **Compact 5** has a centre height of 65mm (2½") and a centre distance of 350mm (13¾"). Powered by a 500W (^{2}/₃hp) motor, it has six speeds between 200 and 2,400rpm. It weighs in at 20kg (36lb).

The **Compact 8E** has a centre height of 105mm (4") and a distance between centres of 450mm (17¾"). It is powered by a 650watt (¾hp) motor giving six speeds between 100 and 1,700rpm. It is available in metric and imperial versions.

The **Ecomat** tool-room range comprises five lathes from the smallest **14D** to the largest **Ecomat 20D**. With centre heights from 140 - 200mm (5½" - 8") and centre distance of 650 - 1,000mm (24²/₃" - 39¹/₃"), all of them are supplied with Camlok chucks. Motors range from 3.2kW (4¼hp) to 7.5kW (10hp) with electronic variable-speed control. All require a three-phase electrical supply. They weigh from 243 - 865kg (536 - 1,907lb).

Myford

The pedigree of current Myford lathes goes back almost 60 years. They comprise the entry-level **Super 7 Sigma Plus**, standard **Super 7 Plus** and **Super 7 Connoisseur**.

All lathes feature a tee-slotted cross slide with a top slide that swivels 360°, back gear, screw cutting and bench

Figure 8. *The Myford Super 7 Connoisseur, Vari-speed, screw-cutting lathe.*

mounting with 89mm (3½") centre height and either 475mm (18¾") or 780mm (30¾") distance between centres. The lathes can all be supplied either in metric or in imperial format.

The **Super 7 Sigma Plus** and **Super 7 Plus** both provide 14 belt-driven speeds from 27 - 2,105rpm (20% faster on US machines) driven by a 0.55kW (¾hp) single-phase motor. A quick-change screw-cutting gearbox is available. The Connoisseur has a variable-speed-control unit with a 0.75kW (1hp), three-phase motor, poly 'V' belts and is ready mounted on an industrial cabinet stand.

The basic lathes are supplied with some accessories included in the price, such as a 3-jaw chuck with fitted back plate, tailstock hard centre, set of change wheels, basic tools, lubricating kit and an installation and user manual. Other options include stands, chucks, steadies, collets, a rack-operated tailstock, rotating centres and tooling sets.

Peatol/Taig

Known in the UK as Peatol and in the US as Taig, this small lathe is light enough to be portable and is suitable for clock making with an overall working accuracy

14

Figure 9. *The Peatol/Taig lathe without a motor fitted. Photo courtesy Peatol Machine Tools.*

Figure 10. *The Proxxon PD 230E lathe. Photo courtesy BriMarc/Proxxon.*

0.0005" (0.0127mm). However, it has no thread-turning capability. It can be supplied complete or as a kit.

The centre height is 2¼" (57mm) and the distance between centres is 9" (228mm) using the optional tailstock as an end support.

The carriage will move 9" (228mm) and there is a cross-slide movement of 1¾" (37mm). The use of raising blocks increases the centre height to 3" (76mm). A ¼hp (200W) electric motor drives the spindle at up to 7,000rpm; the speed range depending on the selected motor and pulleys used.

Accessories include various 3-jaw, 4-jaw and drill chucks, a face plate, vertical and compound slides, a fixed steady, a live centre and a ball-turning tool.

Proxxon

This German company makes two small lathes. The **PD 400** screw-cutting lathe has a centre distance of 400mm (15¾") and a centre height of 85mm (3¹/₃"). Cross-slide travel is 85mm (3¹/₃"); the top slide moves 55mm (2¹/₈"). A 550W (¾hp) reversible motor provides six belt-driven spindle speeds from 80 - 2,800rpm. The overall size is 900 x 400 x 300mm (35⁷/₁₆" x 15¾" x 11¾"). It weighs 45kg (99lb). A 100mm (4") 3-jaw chuck, a quick-change tool post with two holders, a tailstock chuck and a live centre come as standard.

The **PD 230/E** has a centre height of 58mm (2¼") and a centre distance of 230mm (9"). The cross-slide travel is 60mm (2¹/₃") and the top slide moves 45mm (1¾"). A 140W (0.2hp) mains motor provides three belt-driven ratios combined with electronic speed control to provide speeds from 100 - 3,000rpm. Its dimensions are 530mm x 250mm (20⁷/₈" x 9⁷/₈"). It is 150mm (6") high and weighs 10kg (22lb). It includes as standard a 3-jaw chuck and a live centre.

South Bend

In the United States, South Bend lathes have been made for over a century. They are relatively large, quality machine tools with a swing of 14" (356mm) and with distances between centers of 40" to 60" (1,016 to 1,524mm). Unfortunately, they are all big machines that weigh at least 2,970lb (1,350kg) and require three-phase electrical power. There are promises of a 7" x 15" (178 x 381mm) lathe as well as several with 10" (250mm) and 12" (300mm) swings.

Sherline

Sherline manufactures small lathes with a 3½" (89mm) swing and center distances of 8" or 17" (203mm or 431mm). A DC motor gives variable speeds of 70 - 2,800rpm from UK or US mains.

The basic model is the **Model 4000 (metric 4100)** lathe, which includes the

Figure 11. *The Sherline 4000 lathe with a range of accessories. Photo courtesy Sherline.*

Figure 12. The smallest Chinese-made lathe is the Sieg C0 lathe, which weighs just 13kg (29lb). It can be purchased in the UK from Arc Euro Trade or Axminster. Photo courtesy Arc Euro Trade.

motor and its speed controller, a 2¾" x 6" (70 x 152mm) cross slide and a 15" (381mm) steel bed that allows 8" (203mm) between centers. It comes with pulleys, a belt, a faceplate, a lathe dog, two dead centers, two hexagonal keys, a tool post, a high-speed cutting tool and a manual.

The **Model 4500 (metric 4530)** lathe is listed with the same equipment above, but with the addition of larger resettable "zero" hand wheels on both the lead screw and the cross slide.

The **Model 4400 (metric 4410)** lathe is the same as the **Model 4000** but has a 24" (609mm) bed, which provides 17" (431mm) between centers, and a rocker tool post instead of the standard tool post.

Wabeco

The Wabeco lathes are made in Germany. The **D2000 E**, illustrated in Figure 1 on page 10, has a centre distance of 350mm (13¾") and a centre height of 110mm (4"). It measures 1,050 x 420 x 410mm (41¹/₃" x 16½" x 16¹/₈") and weighs 59kg (130lb).

The **D2400 E** and **D3000 E** increase the centre distance to 500mm. The **D2400 E** measures 1,200 x 420 x 410mm (47¼" x 16½" x 16¹/₈"). The **D3000 E** is the same size. They weigh 65kg (143lb) and 71kg (156lb) respectively.

The **D4000 E** has a centre distance of 350mm (13¾") and a centre height of

100mm (4"). It measures 860 x 380 x 400mm (33⁷/₈" x 15" x 15¾") and weighs 71kg (156lb).

The **D6000 E** has a centre distance of 600mm (23½") and a centre height of 135mm (5¹/₃") and measures 1,230 x 470 x 500mm (48½" x 18½" x 19²/₃"). It weighs 150kg (330lb).

The lathes employ electronic infinitely variable speed control with 1.4kW (2hp) motors giving speeds from 30 to 2,300rpm.

The **D6000 E high-speed** lathe has the same centre height and centre distance as the **D6000 E** but measures 1,200 x 479 x 630mm (47¼" x 18⁷/₈" x 24¾") and weighs 177kg (389lb). It has a Camlock chuck and a 2.0kW (2²/₃hp) motor that increases the speed range to 100 - 5,000rpm.

Accessories include a base cabinet, a coolant system, both fixed and moving steadies, live centres, a vertical slide and a 3-axis digital readout system with linear measuring scales.

New Far-East lathes

Many companies offer a wide range of lathes, mainly built in China. While their

Figure 13. *The Warco GH 1440 lathe is typical of large Chinese-made lathes. Photo courtesy Warco.*

quality may not match machines made in Europe or the United States, their highly competitive prices are attractive to many model engineers. Most of these lathes will not face very heavy use and there is little concern about absolute accuracy as home-made parts are easily made to fit.

UK companies like Axminster, Chester, and Warco offer a huge choice of lathes as do many US companies like Grizzly Industrial Inc and Harbor Freight. However, it is worth noting that most of these lathes are made by just a few suppliers like Real Bull Machine Tool, Shanghai Sieg Machinery and Weiss Machinery.

Jet machinery was launched by the Swiss Walter Meier Group in 1970. Their manufacturing is mostly done by Chinese partners. In Germany, Optimum Maschinen offers its Opti and Quantum machines

across Europe. They are imported in the UK by Excel Machine Tools. Since 2003, most of Optimum's machines have been made in its factory in China with German production and quality managers. Their design, development and quality management are undertaken mainly in Germany.

Popular lathes come in many sizes. The centre heights range from just 51mm (2") to 190mm (7½") and centre distances from 125mm (5") to 1,015mm (40") or more. Motors may be small single-phase motors or 2.2kW (3hp) three-phase ones. Weights are from under 20kg (44lb) to 1,000kg (2,200lb) and there are many intermediate sizes from which to choose.

Larger and heavier lathes that weigh more than one tonne (ton) have all been excluded from this book. This is because of their weight and also their size.

Lathe sold in UK	Centre height	Centre distance	Headstock spindle bore	Motor	1 or 3 phase
Axminster Sieg C0	55mm (2⅛")	125mm (5")	10mm ($^2/_5$")	150W ($^1/_5$hp)	1
Chester Cobra	70mm (2¾")	140mm (5¾")	9mm ($^1/_3$")	150W ($^1/_5$hp)	1
Axminster Sieg C1	70mm (2¾")	250mm (9¾")	9mm ($^1/_3$")	150W ($^1/_5$hp)	1
Clarke CLM250M	70mm (2¾")	250mm (9¾")	9mm ($^1/_3$")	150W ($^1/_5$hp)	1
Quantum D140 x 250 Vario	70mm (2¾")	250mm (9¾")	11mm (½")	450W ($^2/_3$hp)	1
Axminster Sieg C2A	90mm (3½")	300mm (12")	20mm (¾")	250W ($^1/_3$hp)	1
Clarke CLM300M	90mm (3½")	300mm (12")	20mm (¾")	300W ($^2/_5$hp)	1
Chester DB7VS	90mm (3½")	300mm (12")	21mm ($^7/_8$")	700W (1hp)	1
Opti D 180 x 300 Vario	90mm (3½")	300mm (12")	21mm ($^7/_8$")	600W ($^4/_5$hp)	1
Warco WM-180	90mm (3½")	300mm (12")	21mm ($^7/_8$")	550W (¾hp)	1
Warco Mini Lathe	90mm (3½")	300mm (12")	20mm (¾")	550W (¾hp)	1
Chester Conquest	90mm (3½")	325mm (12¾")	19mm (¾")	400W (½hp)	1
Chester Conquest Super	90mm (3½")	325mm (12¾")	19mm (¾")	400W (½hp)	1
Arc Euro Trade C3	90mm (3½")	350mm (14")	20mm (¾")	350W (½hp)	1

Table 1a. *The range of Far-East-manufactured lathes sold in the UK, in order of centre-height.*

Table 1 shows the range of lathes that are made in the Far East and imported to the UK. They were all being offered on the market early in 2010.

They are listed first in order of centre heights and then by their distance between centres. Information about the spindle bore, motor power, electrical supply and speed range, overall dimensions and total weight are also listed.

Figure 14. *The Clarke CLM300M lathe. Photo courtesy Chronos.*

The **Axminster Sieg C0 Metal Turning Lathe**, which is also available from **Arc Euro Trade**, is the smallest of all these lathes and comes with a 50mm (2") 3-jaw chuck, a single tool post suitable for 8mm ($^1/_3$") tools, a splash guard and an interlocked safety chuck guard.

The **Chester Cobra** is only marginally larger than the **Sieg C0** and includes an 80mm (3") 3-jaw chuck, a 2-way tool post, a steel centre and a set of basic tools.

The **Clarke CL250M Metal Lathe** is fitted with a 3-jaw chuck, a single tool post, a fixed centre and a set of tools. Its specification, which is virtually identical to the **Axminster Sieg C1 Micro Lathe Mk2**, also available from Arc Euro Trade, includes an 80mm (3") 3-jaw chuck, a tailstock centre and a set of service tools.

The **Quantum D140 x 250 Vario** is made by Optimum Maschinen and is available in the UK from Excel Machine Tools. It comes with an 80mm (3") 3-jaw chuck, a 4-way tool post, a pair of fixed centres, a splash guard and a chip pan.

Speeds		Dimensions length x width x height	Weight
100 - 3,850	VS	440 x 270 x 210mm (17 1/3" x 10 2/3" x 8¼")	13kg (29lb)
100 - 2,000	VS	540 x 300 x 270mm (21½" x 12" x 11")	23kg (51lb)
100 - 2,000	VS	630 x 330 x 210mm (24¾" x 13" x 8¼")	22kg (48lb)
100 - 2,000	VS	630 x 330 x 210mm (24¾" x 13" x 8¼")	23kg (51lb)
120 - 3,000	VS	550 x 320 x 260mm (21 2/3" x 12 2/3" x 10¼")	19kg (42lb)
100 - 2,550	2VS	720 x 300 x 290mm (28 1/3" x 11¾" x 11½")	37kg (81lb)
100 - 2,500	VS	820 x 295 x 300mm (32¼" x 11 2/3" x 11¾")	40kg (88lb)
50 - 2,500	VS	780 x 480 x 420mm (30¾" x 18 7/8" x 16½")	55kg (121lb)
150 - 2,500	2VS	830 x 396 x 355mm (32 2/3" x 15 5/8" x 14")	55kg (121lb)
0 - 2,500	2VS	813 x 380 x 330mm (32" x 15" x 13")	70kg (154lb)
50 - 2,900	VS	700 x 254 x 260mm (28" x 10" x 10")	38kg (84lb)
0 - 2,500	VS	770 x 254 x 300mm (30 1/3" x 10" x 12")	34kg (74lb)
0 - 2,500	VS	771 x 254 x 300mm (30 1/3" x 10" x 12")	38kg (84lb)
100 - 3,000	2VS	750 x 320 x 330mm (29½" x 12½" x 13")	44kg (97lb)

VS = variable speed, 2VS = two separate variable speed ranges.

The next nine lathes, comprising the **Axminster Sieg C2A**, the **Clarke CLM300M**, the **Chester DB7VS**, the **Opti D 180 x 300 Vario**, the **Warco WM-180**, the **Warco Mini Lathe**, the **Chester Conquest** and **Super Conquest** and the **Arc Euro Trade C3**, all have the 90mm (3½") centre height that is so popular in the United Kingdom. They are all supplied complete with a minimum of an 80mm (3") 3-jaw chuck, a tail-stock centre and a set of service tools.

The **Warco Mini Lathe**, the **Clarke CLM300M** and **Chester Conquest** all have a 4-way tool post as well. The **Chester DB7VS** and **Warco WM-180** also have a 4-jaw chuck, faceplate and fixed and travelling steadies while the **Chester Conquest Super** and **Arc Euro Trade C3** additionally come with a dead centre and a digital display of spindle-speed. Their sizes, weights and motor power do not vary enormously and will depend on the particular lathe being considered.

Figure 15. *The Sieg C1 lathe is available both from Axminster and Arc Euro Trade.*

Figure 16. *The Opti D 180 x 300 Vario lathe. Photo courtesy Optimum Machinen.*

Lathe sold in UK	Centre height	Centre distance	Headstock spindle bore	Motor	1 or 3 phase
Warco BV-20	100mm (4")	350mm (14")	20mm (¾")	550W (¾hp)	1
Warco WM-240	105mm (4")	400mm (15¾")	21mm (⁷/₈")	600W (⁴/₅hp)	1
Chester DB8	105mm (4")	400mm (15¾")	20mm (¾")	550W (¾hp)	1
Chester DB8VS	105mm (4")	400mm (15¾")	21mm (⁷/₈")	750W (1hp)	1
Chester Comet	105mm (4")	450mm (17¾")	20mm (¾")	1.1kW (1½hp)	1
Quantum D 210 x 400	105mm (4")	400mm (15¾")	21mm (⁷/₈")	750W (1hp)	1
Axminster Sieg C4	109mm (4⅓")	410mm (16")	20mm (¾")	1kW (1⅓hp)	1
Axminster BV20M2	110mm (4⅓")	350mm (14")	20mm (¾")	375W (½hp)	1
Axminster Jet BD920W	110mm (4⅓")	500mm (19½")	20mm (¾")	550W (¾hp)	1
Warco 918	114mm (4½")	482mm (19")	19mm (¾")	550W (¾hp)	1
Chester 920	120mm (4¾")	500mm (19½")	19mm (¾")	550W (¾hp)	1
Quantum D 250 x 400	125mm (5")	450mm (17¾")	21mm (⁷/₈")	750W (1hp)	1

Table 1b. *The range of Far-East-manufactured lathes sold in the UK, in order of centre-height.*

The following six lathes all have exactly the same centre height but the various other parameters do show only some fairly minor variations.

The **Warco BV-20** has six speeds that are selected through a gearbox, the **Warco WM-240** has two variable-speed ranges, the **Chester DB8** has six belt-driven speeds while the **DB8VS** is a variable-speed lathe. All four are similarly equipped

Figure 17. *The Warco BV-20 lathe. Photo courtesy Warco.*

with 3- and 4-jaw chucks, a faceplate, fixed and travelling steadies, a 4-way tool post, two dead centres, a swarf tray and a rear chip guard.

The **Chester Comet** is a variable speed lathe that is driven by a significantly more powerful motor. It comes with a 3-jaw chuck, a faceplate, both fixed and travelling steadies, a quick-change tool post, two dead centres, a swarf tray and a rear chip guard.

The **Quantum D 210 x 400 G** has a 100mm (4") 3-jaw chuck, a 4-way tool post, two fixed centres, a lathe dog, a rear splash guard, a chip pan and a set of tools. The speeds of the **Vario** version are infinitely variable from 150 - 2,200rpm.

Table 1b continues with a further six lathes, which all have centre heights of between 109mm (4¼") and 120mm (4¾"). Their range of distances between centres vary slightly from 410mm (16") to 500mm (19½").

The **Axminster Sieg C4** features a touch-screen control panel and a digital speed readout as well as cross-slide and longitudinal power feeds. It comes with

Speeds		Dimensions length x width x height	Weight
140 - 1,710	6g	1.016 x 483 x 400mm (40" x 19" x 16")	105kg (231lb)
50 - 2,200	2VS	1,066 x 585 x 430mm (42" x 23" x 17")	110kg (242lb)
125 - 2,000	6	860 x 360 x 548mm (33^7/$_8$" x 14" x 21½")	125kg (275lb)
50 - 2,000	VS	1,050 x 560 x 570mm (41^1/$_3$" x 22" x 22½")	110kg (242lb)
100 - 2,000	VS	1,000 x 550 x 400mm (39^1/$_3$" x 21^2/$_3$" x 15½")	94kg (207lb)
120 - 2,000	6	880 x 500 x 475mm (34^2/$_3$" x 19^2/$_3$" x 18¾")	80kg (176lb)
50 - 2,000	VS	1,000 x 550 x 400mm (39^1/$_3$" x 21^2/$_3$" x 15¾")	125kg (275lb)
170 - 1,950	6g	915 x 650 x 560mm (36" x 25½" x 22")	140kg (308lb)
110 - 1,700	6g	955 x 508 x 1,143mm (37½" x 20" x 45")	110kg (242lb)
100 - 1,800	6g	955 x 485 x 393mm (37½" x 19" x 15½")	98kg (215lb)
100 - 1,800	6	1,040 x 540 x 380mm (41" x 21¼" x 15")	100kg (220lb)
125 - 2,000	6	865 x 500 x 500mm (34" x 19^2/$_3$" x 19^2/$_3$")	96kg (211lb)

VS = variable speed, 2VS = two separate variable speed ranges, g = a gearbox.

a 3-jaw chuck and a 4-way tool post. Two options are a stand and a milling head.

The **Axminster BV20M2** has a 100mm (4") 3-jaw chuck, a 4-way tool post, a tailstock centre and service tools. Options include a floor stand, a 100mm (4") 4-jaw chuck, a pair of steadies and a faceplate.

The **Axminster Jet BD 920 W** comes with a 3-jaw self-centring chuck, a 4-way tool post, fixed and travelling steadies as well as an optional floor stand.

The **Warco 918** lathe with quick-change gearbox was introduced as long ago as 1985 by Warco. It includes 100mm (4") 3-jaw and 4-jaw chucks, a 4-way tool post, a faceplate, fixed and travelling steadies, two dead centres and a set of tools. Its options include a floor stand, a quick-change tool post, a collet chuck, a tailstock die holder and a live centre.

The **Chester 920** includes 100mm (4") 3-jaw and 125mm (5") 4-jaw chucks, a 4-way tool post, a 190mm (7½") faceplate, fixed and travelling steadies, two dead centres, a rear splash guard, a machine tray and a set of tools. Options include a floor stand, a coolant system, a vertical

slide, a flexi-arm lamp, a drill chuck and arbor and a live centre.

The **Quantum D 250 x 400** is the smallest of the 125mm (5") centre-height lathes with a distance between centres of just 450mm (17¾"). It is also available in **Vario** form with electronic speed control from 20 - 2,200rpm. Both machines come complete with a 125mm (5") 3-jaw chuck, a 4-way tool post, two fixed centres, a lathe dog, a rear splash guard, a chip pan and a set of tools.

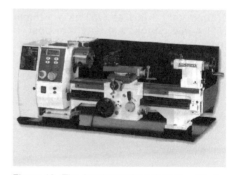

Figure 18. *The Axminster Sieg C4 lathe. Photo courtesy Axminster Power Tool Centre.*

Lathe sold in UK	Centre height	Centre distance	Headstock spindle bore	Motor	1 or 3 phase
Warco WM-250	125mm (5")	550mm (22")	26mm (1")	750W (1hp)	1
Axminster Sieg C6B	125mm (5")	550mm (22")	20mm (¾")	800W (1hp)	1
Chester DB10G	125mm (5")	550mm (22")	20mm (¾")	550W (¾hp)	1
Chester DB10GVS	125mm (5")	550mm (22")	26mm (1")	1.2kW (1⅔hp)	1
Opti D 240 x 500 G	125mm (5")	550mm (22")	26mm (1")	750W (1hp)	1
Quantum D 250 x 550	125mm (5")	550mm (22")	21mm (⅞")	750W (1hp)	1
Chester DB11GVS	140mm (5½")	700mm (27½")	26mm (1")	1.2kW (1⅔hp)	1
Warco WM-280	140mm (5½")	700mm (27½")	26mm (1")	1.1kW (1½hp)	1
Warco WM-280V-F	140mm (5½")	700mm (27½")	26mm (1")	1.1kW (1½hp)	1
Opti D 280 x 700 G	140mm (5½")	700mm (27½")	26mm (1")	850W (1⅛hp)	1

Table 1c. *The range of Far-East-manufactured lathes sold in the UK, in order of centre-height.*

Next are six 125mm (5") centre-height lathes with a 550mm (22") centre distance.

The **Warco WM-250** variable-speed lathe is provided with 3- and 4-jaw chucks, a 4-way tool post, fixed and travelling steadies, two dead centres, a rear splash guard and a swarf tray.

The **Axminster Sieg C6B** includes an electronic speed control, a 125mm (5") 3-jaw chuck, a 4-way indexing tool post and an integral rear-splash guard. Options include a 4-jaw chuck, both fixed and travelling steadies, a 220mm (8⅔") faceplate and a cabinet stand.

The **Chester DB10G** comes with a 100mm (4") 3-jaw as well as a 125mm (5") 4-jaw chuck, a 4-way tool post, a faceplate, fixed and travelling steadies,

Figure 19. *The Quantum D 250 x 550 lathe. Photo courtesy Optimum Machinen.*

two dead centres, a set of lathe tools, a machine tray and a rear splash guard.

The **Chester DB10GVS** comes with the same items but the 3-jaw chuck is 125mm (5"). Accessories for both these Chester lathes include a rotating centre, a quick-change tool post, a vertical slide, a drill chuck, a coolant system and a floor stand.

The **Opti D 240 x 500 G** has a 125mm (5") 3-jaw chuck, a 4-way tool post, two fixed centres, a lathe dog and a set of tools. The **Vario** is only slightly different with a 1.5kW (2hp) motor and four variable speed ranges from 30 - 4,000rpm. It is 90mm (3½") wider.

The **Quantum D 250 x 550**, made by the same company, is very similar and also can be purchased in **Vario** form with electronic-speed control from 20 - 2,500rpm. Both variants come with a 125mm (5") 3-jaw chuck, a 4-way tool post, two fixed centres, a lathe dog, a rear splash guard, a chip pan and a set of tools. The last four lathes have centre heights of 140mm (5½"), centre distances of 700mm (27½") and similar weights.

The **Chester DB11GVS** has 125mm (5") 3-jaw and 4-jaw chucks, a face plate, a 4-way tool post, both fixed and travelling

Speeds		Dimensions length x width x height	Weight
50 - 2,500	2VS	1,194 x 610 x 432mm (47" x 24" x 17")	120kg (264lb)
100 - 2,000	VS	1,200 x 660 x 600mm (47¼" x 26" x 23²/₃")	145kg (319lb)
130 - 2,100	6	1,015 x 548 x 500mm (40" x 21½" x 19²/₃")	150kg (330lb)
50 - 2,500	VS	1,110 x 580 x 480mm (43¾" x 22¾" x 19")	120kg (264lb)
125 - 2,000	6	1,250 x 585 x 475mm (49¼" x 23" x 18¾")	125kg (275lb)
125 - 2,000	6	1,015 x 500 x 500mm (40" x 19²/₃" x 19²/₃")	125kg (275lb)
125 - 2,500	VS	1,390 x 700 x 630mm (54¾" x 27½" x 24¾")	180kg (396lb)
50 - 2,500	2VS	1,302 x 635 x 508mm (52" x 25" x 20")	190kg (418lb)
125 - 2,000	VS	1,302 x 635 x 508mm (52" x 25" x 20")	190kg (418lb)
150 - 2,000	6	1,370 x 640 x 535mm (54" x 25¼" x 21")	180kg (396lb)

VS = variable speed, 2VS = two separate variable speed ranges.

steadies, two fixed centres, a set of lathe tools, a machine tray, a rear splash guard and a chip guard. Options include a stand, a coolant system, a light, a live centre, a vertical slide, a quick-change tool post, a drill chuck and a set of lathe tools.

The two **Warco WM-280** lathes in this last category are very similar. The basic **WM-280** has two gears giving a range of ·variable speeds. The **WM-280V-F** uses a single range of variable speeds and has a dedicated shaft for longitudinal and cross feed. Both lathes come with 3-jaw and 4-jaw chucks, a faceplate, a 4-way

tool post, fixed and travelling steadies, two dead centres, a swarf tray and a rear splash guard. The wide range of options includes a stand, a quick-change tool post, a collet chuck, a live centre, a drill chuck and a tailstock die holder.

The **Opti D 280 x 700 G** comes with a 125mm (5") 3-jaw chuck, two centres, a lathe dog, a pair of steadies, a 4-way tool post and HSS cutting tools, a splash board, a chip tray and a set of tools. The **Vario** has a more powerful 1.5kW (2hp) motor combined with four infinitely variable speed ranges.

Figure 20. *The Warco WM-280V-F variable-speed lathe. Photo courtesy Warco.*

Lathe sold in UK	Centre height	Centre distance	Headstock spindle bore	Motor	1 or 3 phase
Clarke CLM430	152mm (6")	430mm (17")	38mm (1½")	550W (¾hp)	1
Warco WMT300	150mm (6")	500mm (19½")	26mm (1")	550W (¾hp)	1
Chester Cub 620*	150mm (6")	500mm (19½")	38mm (1½")	1.1kW (1½hp)	1
Chester Craftsman*	150mm (6")	570mm (22½")	36mm ($1^2/_5$")	1.1kW (1½hp)	1
Warco BH600*	150mm (6")	610mm (24")	36mm ($1^2/_5$")	1.5kW (2hp)	1
Warco GH/VS 1224*	150mm (6")	635mm (25")	38mm (1½")	1.5kW (2hp)	1
Axminster BV30M	150mm (6")	650mm ($25^2/_3$")	38mm (1½")	750W (1hp)	1
Chester Cub 630*	150mm (6")	750mm (29½")	38mm (1½")	1.1kW (1½hp)	1
Chester Crusader *	150mm (6")	810mm (32")	38mm (1½")	1.1kW (1½hp)	1
Axminster Sieg CQ6230B*	150mm (6")	900mm (35½")	38mm (1½")	1.5kW (2hp)	1
Warco GH/VS1232*	150mm (6")	914mm (36")	38mm (1½")	1.5kW (2hp)	1
Warco BH900*	150mm (6")	940mm (37")	36mm ($1^2/_5$")	1.5kW (2hp)	1

Table 1d. *The range of Far-East-manufactured lathes sold in the UK, in order of centre-height.*

Far East lathes with a 152mm (6") centre height are numerous and come with a range of distances between centres from 430 - 940mm (17" - 37").

The **Clarke CLM430** is fitted with a 3-jaw chuck, a 4-way tool post, two fixed centres, a machine vice and a set of tools. An optional stand complete with suds tray can also be purchased.

The **Warco WMT300** has a 125mm (5") 3-jaw chuck and a 200mm (8") faceplate, a 4-way tool post, fixed and travelling steadies, two fixed centres, a 12.5mm (½") drill chuck and a set of tungsten-carbide-tipped cutting tools. The normal range of Warco accessories includes a stand, a quick-change tool post, a collet chuck, a live centre, a drill chuck, a tailstock die holder and a range of indexable lathe tools.

The **Chester Cub 620** and **Cub 630** are identical apart from their centre distances, overall lengths and their total weights. They come with a 150mm (6") 3-jaw chuck, a 200mm (8") 4-jaw chuck, a 290mm (11½") faceplate, a 4-way tool post, fixed and travelling steadies, a live centre, a built-in coolant system, a stand with rear splash guard and a low-voltage halogen light. Options include a quick-change tool post, a vertical slide and a drill chuck.

The **Chester Craftsman** has 160mm (6⅓") 3-jaw and 4-jaw chucks and a faceplate. It also has a 4-way tool post, fixed and travelling steadies, two fixed centres and a stand with a splash guard. In addition to all the **Cub** options, a coolant system and lamp are also available for the **Craftsman**.

Figure 21. *The Chester Craftsman lathe.*

Speeds		Dimensions length x width x height	Weight
170-1,630	6	1,100 x 600 x 405mm (43⅓" x 23⅔" x 16")	129kg (284lb)
160 - 1,600	6	950 x 580 x 410mm (37½" x 22¾" x 16")	140kg (308lb)
60 - 2,000	9g	1,400 x 680 x 1,200mm (55" x 26¾" x 47¼")	480kg (1,056lb)
50 - 1,200	12	1,370 x 740 x 680mm (54" x 29" x 26¾")	390kg (858lb)
50 - 1,200	12	1,300 x 686 x 1,245mm (51" x 27" x 49")	360kg (795lb)
75 - 1,400	9g/VS	1,320 x 610 x 1,250mm (52" x 24" x 49")	550kg (1,210lb)
75 - 1,450	9g	1,450 x 680 x 450mm (57" x 26¾" x 17¾")	280kg (616lb)
60 - 2,000	9g	1,650 x 680 x 1,200mm (65" x 26¾" x 47¼")	500kg (1,100lb)
65 - 1,810	18g	1,780 x 650 x 1,160mm (70" x 25½" x 45⅔")	450kg (990lb)
75 - 1,400	g	1,680 x 750 x 1,400mm (66" x 29½" x 55")	530kg (1,186lb)
75 - 1,400	9g/VS	1,524 x 610 x 1,250mm (60" x 24" x 49")	560kg (1,232lb)
50 - 1,200	12	1,630 x 686 x 1,245mm (64" x 27" x 49")	410kg (902lb)

* = a gap bed, VS = variable speed, g = a gearbox.

There are seven 152mm (6") centre height Warco lathes including the **WMT300**.

The **Warco BH600** and **BH900** are identical apart from their distance between centres, overall length and weight. The **Warco GH1224** and **GH1232** have nine gears to provide different speeds, while both the **Warco VS1224** and **VS1232** have an almost infinitely variable speed range. The major differences between the **GH/VS1224** and **GH/VS1232** are their centre distances, total lengths and weights. These six lathes include a cabinet stand, 3-jaw and 4-jaw chucks, a faceplate, a 4-way tool post, fixed and travelling steadies and two dead centres. Both pairs of **1224** and **1232** also have a halogen low-voltage light.

The **Axminster BV30M** is supplied complete with a 3-jaw chuck, headstock and tailstock centres and a set of service tools. An optional floor stand is available. Other options include a 4-jaw independent chuck, a choice of two faceplates and a travelling steady.

The **Chester Crusader** comes with a similar set of standard items to the **Cub** but also has a quick-change tool post and a foot switch. It has an almost identical set of optional extras.

The **Axminster Sieg CQ6230B** has 150mm (6") 3- and 4-jaw chucks, a quick-change tool post, a power cross feed, live and dead centres, a built-in floor stand with oil tray, a foot brake and a coolant system, together with a work lamp and a set of tools.

Figure 22. *The Warco BH600 lathe. Photo courtesy Warco.*

25

Lathe sold in UK	Centre height	Centre distance	Headstock spindle bore	Motor	1 or 3 phase
Warco GH1322*	165mm (6½")	500mm (19½")	38mm (1½")	1.5kW (2hp)	1/3
Chester Coventry 500*	166mm (6½")	500mm (19½")	38mm (1½")	1.5kW (2hp)	1/3
Warco GH 550	165mm (6½")	600mm (23½")	38mm (1½")	1.1kW (1½hp)	1
Opti D 320 x 630*	160mm (6⅓")	630mm (24¾")	38mm (1½")	1.5kW (2hp)	3
Chester Coventry 750*	166mm (6½")	750mm (29½")	38mm (1½")	1.5kW (2hp)	1/3
Warco GH1330*	165mm (6½")	920mm (36¼")	38mm (1½")	1.5kW (2hp)	1/3
Opti D 320 x 920*	160mm (6⅓")	920mm (36¼")	38mm (1½")	1.5kW (2hp)	3
Axminster Runmaster	166mm (6½")	1,000mm (39⅓")	38mm (1½")	2.2kW (3hp)	1/3
Chester Coventry 1000*	166mm (6½")	1,000mm (39⅓")	38mm (1½")	1.5kW (2hp)	1/3
Warco GH1339*	165mm (6½")	1,000mm (39⅓")	38mm (1½")	1.5kW (2hp)	1/3
Opti D 330 x 1000*	165mm (6½")	1,000mm (39⅓")	38mm (1½")	1.5kW (2hp)	3
Axminster Jet GHB1340A	165mm (6½")	1,015mm (40")	38mm (1½")	1.4kW (2hp)	3
Chester Challenger 750*	175mm (7")	750mm (29½")	38mm (1½")	2.2kW (3hp)	3
Chester Challenger 1000*	175mm (7")	1,000mm (39⅓")	38mm (1½")	2.2kW (3hp)	3
Axminster Jet GH1440W*	178mm (7")	1,015mm (40")	38mm (1½")	2.2kW (3hp)	3
Opti D 360 x 1000*	180mm (7")	1,000mm (39⅓")	38mm (1½")	2.4kW (3¼hp)	3
Warco GH-1440*	190mm (7½")	1,000mm (39⅔")	38mm (1½")	2.2kW (3hp)	3

Table 1e. *The range of Far-East-manufactured lathes sold in the UK, in order of centre-height.*

The **Warco GH1322** has 150mm (6") 3-jaw and 4-jaw chucks, a faceplate, a 4-way tool post, both fixed and travelling steadies, two dead centres, a cabinet stand, a coolant system and a light.

The **Warco GH1330** and **GH1339** have the same ancillaries but vary in centre distance, length and weight from the **GH1322**.

The **Warco GH 550** is lighter with more spindle speeds. It includes 160mm (6⅓") 3-jaw and 4-jaw chucks, a 250mm (10") faceplate, a 4-way indexing tool post, two steadies, a dead centre, a light and a stand.

The **Chester Coventry 500, 750** and **1000** differ in their centre distances, lengths and weights. All have 3- and 4-jaw chucks, faceplates, 4-way tool posts, pairs of steadies, fixed centres, floor stands, coolant systems and tool boxes.

The **Opti D 320 x 630** and **D 320 x 920** differ only in centre distances, lengths and

weights. Both have Camlok 3-jaw 160mm (6⅓") and 4-jaw 200mm (8") chucks, 250mm (10") face plates, 4-way tool posts, revolving MT3 centres, fixed and moving steadies, lamps, stands and sets of tools.

The **Axminster Runmaster** includes a 150mm (6") 3-jaw Camlok chuck, a 4-way tool post, fixed and travelling steadies, live and dead centres, a light, a coolant system, a cabinet stand and a toolbox.

The **Opti D 330 x 1000** has Camlok 3-jaw 160mm (6⅓") and 4-jaw 200mm (8") chucks, a 250mm (10") face plate, a 4-way tool post, two revolving centres, two steadies, a stand, a lamp and a set of tools.

The **Axminster Jet GHB1340A** has a digital readout showing cross-slide, saddle and top-slide travel. It has a 150mm (6") 3-jaw and a 200mm (8") 4-jaw chuck, a 300mm (12") faceplate, a 4-way tool post, two steadies, a suds tray and a stand.

Speeds	Dimensions length x width x height	Weight
70 - 2,000 8g	1,320 x 610 x 1,400mm (52" x 24" x 55")	500kg (1,110lb)
70 - 2,000 8g	1,350 x 750 x 1,230mm (53" x 29½" x 48½")	400kg (880lb)
60 - 1,650 12g	1,390 x 700 x 1,200mm (54¾" x 27½" x 47¼")	280kg (616lb)
65 - 1,800 18	1,390 x 760 x 1,420mm (54¾" x 30" x 56")	400kg (880lb)
70 - 2,000 8g	1,600 x 750 x 1,230mm (63" x 29½" x 48½")	450kg (990lb)
70 - 2,000 8g	1,524 x 610 x 1,400mm (60" x 24" x 55")	600kg (1,320lb)
65 - 1,800 18	1,680 x 760 x 1,420mm (66⅛" x 30" x 56")	410kg (902lb)
70 - 2,000 8g	1,940 x 755 x 1,238mm (76⅓" x 29¾" x 48¾")	610kg (1,342lb)
70 - 2,000 8g	1,800 x 750 x 1,230mm (70⅞" x 29½" x 48½")	500kg (1,100lb)
70 - 2,000 8g	1,800 x 610 x 1,400mm (70⅞" x 24" x 55")	525kg (1,150lb)
70 - 2,000 8g	1,960 x 755 x 1,400mm (77⅛" x 29¾" x 55")	800kg (1,760lb)
70 - 2,000 8g	1,905 x 762 x 1,200mm (75" x 30" x 47¼")	610kg (1,342lb)
45 - 1,800 16g	1,890 x 760 x 1,200mm (77½" x 30" x 47¼")	750kg (1,650lb)
45 - 1,800 16g	2,140 x 760 x 1,200mm (84¼" x 30" x 47¼")	750kg (1,650lb)
40 - 1,800 12g	1,873 x 750 x 1,200mm (73¾" x 29½" x 47¼")	995kg (2,189lb)
45 - 1,800 16g	1,930 x 760 x 1,580mm (76" x 30" x 62¼")	880kg (1,936lb)
45 - 1,800 16g	1,822 x 735 x 1,235mm (71¾" x 29" x 48½")	750kg (1,650lb)

* = a gap bed, g = a gearbox.

Five lathes with a 175mm (7") centre height or more need a three-phase supply.

The **Chester Challenger 750** and **1000** differ in centre distances and lengths. They have 200mm (8") 3-jaw and 4-jaw Camlok chucks and 300mm (12") faceplates, two steadies, live centres, two-axis readouts, coolant systems, floor cabinets and lights.

The **Axminster Jet GH1440W** is a heavy lathe and has a stand, 160mm (6") 3-jaw and 200mm (8") 4-jaw chucks, a 300mm (12") faceplate, a 4-way tool post, a work light and a digital display of saddle, cross-slide and top-slide travel.

The **Opti D 360 x 1000** has a floor stand with coolant system, a Camlok 3-jaw 160mm (6⅓") and a 4-jaw 200mm (8") chuck, a 260mm (10¼") face plate, a 4-way tool post, an MT3 centre with adaptor, fixed and moving steadies, a chip pan, a machine lamp and a set of tools.

The **Warco GH-1440** comes with a 150mm (6") 3-jaw and a 200mm (8") 4-jaw chuck, a 300mm (12") faceplate, a 4-way tool post, fixed and travelling steadies, dead centres, a coolant system, a floor stand and a low-voltage light.

Figure 23. *The Chester Challenger is a large lathe. Photo courtesy Chester.*

Lathe sold in US	Swing	Centre distance	Headstock spindle bore	Motor	1 or 3 phase
Harbor Fr 95012-4VGA	4" (102mm)	5" (125mm)	³/₈" (10mm)	⅕hp (150W)	1
Harbor Fr 93212-9VGA	7" (180mm)	10" (250mm)	¾" (19mm)	⅓hp (250W)	1
Bolton CQ9318	7" (180mm)	12" (300mm)	¾" (19mm)	¾hp (550W)	1
Grizzly G8688	7" (180mm)	12" (300mm)	¾" (19mm)	¾hp (550W)	1
Harbor Fr 93799-3VGA	7" (180mm)	12" (300mm)	¾" (19mm)	¾hp (550W)	1
Harbor Fr 44859-9VGA	8" (200mm)	12" (300mm)	¾" (19mm)	¾hp (550W)	1
LatheMaster CQ6120X320	8" (200mm)	14" (356mm)	¾" (19mm)	¾hp (550W)	1
Grizzly G4000	9" (228mm)	19" (482mm)	¾" (19mm)	¾hp (550W)	1
Jet BD-920W	9" (228mm)	20" (508mm)	¾" (19mm)	¾hp (550W)	1
Harbor Fr 45861-2VGA	9" (228mm)	20" (508mm)	¾" (19mm)	¾hp (550W)	1
Birmingham YCL-920BD	9" (228mm)	20" (508mm)	1½" (38mm)	¾hp (550W)	1
LatheMaster HD250X750	9" (228mm)	30" (762mm)	¾" (19mm)	¾hp (550W)	1
Bolton CQ9325A	10" (250mm)	15¾" (400mm)	1" (25mm)	¾hp (550W)	1
Grizzly G9972Z	11" (280mm)	26" (660mm)	1" (25mm)	1hp (750W)	1
Birmingham YCL-1126BD	11" (280mm)	36" (914mm)	1½" (38mm)	1hp (750W)	1

Table 2a. *The range of Far-East-built lathes sold in the US, in order of swing.*

Far East lathes in the United States

Table 2 lists 37 lathes imported into the US. These machines are very similar to those imported into the UK but suit US mains electricity. They range in swing from 4" (102mm) to 14" (356mm). In this section, there is American spelling and imperial units precede metric equivalents.

As examples, the **Sears Jet BD-920W** and **GHB-1340A** appear to be the same machines as the **Axminster Jet BD 920 W** and **GHB1340A**. The same applies to the **Grizzly G9249** and the **Warco BH900**. The American market offers fewer small lathes and is oriented towards larger ones, although for this book, the weight limit of 2,200lb (1,000kg) still applies.

The **Harbor Freight 95012-4VGA Micro lathe** includes a 3-jaw chuck, a tool post and a splash guard. It is readily moved onto any suitable workbench.

The larger **Harbor Freight 93212-9VGA Mini lathe** comes with a 3-jaw chuck, a 4-way tool post, a splash guard and a chip tray.

The next three 7 x 12 lathes, the **Bolton CQ9318**, **Grizzly G8688** and **Harbor Freight 93799-3VGA** all have similar specifications and sizes. They all come with a 3-jaw chuck, a 4-way tool post, a splash guard and a chip tray. However, the **Bolton CQ9318** only has one speed range but includes a dead center and a set of Allen keys. The **Grizzly G8688** also has a 6¼" (159mm) face plate, a steady rest, a dead center and a set of wrenches.

The **Harbor Freight 44859-9VGA** has a 4" (100mm) 3-jaw chuck, MT2 and MT3 dead centers, a set of tools and two spare belts.

The **LatheMaster CQ6120X320** comes with both 3 and 4-jaw chucks, a faceplate, a 4-way tool post, two MT dead centers and steady and moving rests.

The **Grizzly G4000** has a 4" (100mm) 3-jaw chuck, a 7¼" (184mm) 4-jaw chuck, steady and moving rests, two dead and a

Speeds		Dimensions length x width x height	Weight
0 - 3,800	VS	14" x 3⁵/₁₆" x 5" (356 x 84 x 127mm)	36lb (16kg)
200 - 3,000	2VS	26¾" x 14" x 13¹/₃" (680 x 356 x 338mm)	89lb (40kg)
200 - 2,500	VS	31" x 12¼" x 13¼" (785 x 538 x 335mm)	80lb (36kg)
0 - 2,500	2VS	27¾" x 11½" x 12" (705 x 292 x 305mm)	74lb (34kg)
186 - 3,000	2VS	30¹/₃" x 13" x 13" (770 x 330 x 330mm)	89lb (40kg)
125 - 2,000	6	37" x 22" x 21¹/₃" (940 x 559 x 541mm)	254lb (115kg)
125 - 2,000	6	34" x 17½" x 17½" (836 x 445 x 445mm)	190lb (86kg)
130 - 2,000	6	37" x 20" x 15" (940 x 508 x 301mm)	250lb (114kg)
130 - 2,000	6	42" x 22" x 21" (1,067mm x 559mm x 533mm)	235lb (107kg)
120 - 2,000	6	40" x 21¹/₃" x 20¹/₃" (1,016 x 541 x 516mm)	230lb (105kg)
120 - 2,000	6	38" x 20" x 19" (965 x 508 x 483mm)	248lb (113kg)
125 - 2,000	6	56" x 19" x 18½" (1,422 x 483 x 470mm)	330lb (150kg)
125 - 1,800	9g	41¹/₃" x 18½" x 13¹/₃" (1,050 x 470 x 340mm)	264lb (120kg)
150 - 2,400	6	51" x 23" x 19" (1,295 x 584 x 483mm)	490lb (223kg)
150 - 2,400	6	52" x 28" x 23" (1,321 x 711 x 584mm)	368lb (167kg)

VS = variable speed, 2VS = two separate speed ranges, g = a gearbox.

live center, a 4-way and C-type tool post and a tool box with tools.

The **Jet BD-920W** comes with 4" (100mm) 3-jaw and a 7" (179mm) 4-jaw chuck, a faceplate, single and 4-way tool posts, steady and moving rests, centers and a toolbox with tools.

The **Harbor Freight 45861-2VGA** has a 4" (100mm) 3-jaw chuck, two dead and a live center, two drive belts and some tools.

The **Birmingham YCL-920BD** comes with a 4" (100mm) 3-jaw and a 7" (179mm) 4-jaw chuck, an 8" (200mm) face plate, steady and moving rests, single and 4-way tool posts, two dead centers, a splash guard and a tool box. There is an optional stand.

The **LatheMaster HD250X750** includes 5" (127mm) 3- and 4-jaw chucks, a faceplate, a 4-way tool post, two dead centers and steady and travelling rests.

The **Bolton CQ9325A** has only a 3-jaw chuck, two MT dead centers and a set of tools but many options are available.

The **Grizzly G9972Z** comes with a 5" (127mm) 3-jaw and a 6½" (165mm) 4-jaw chuck, a faceplate, a 4-way tool post, steady and travelling rests, two MT dead centers and a toolbox with tools.

The **Birmingham YCL-1126BD** comes with a 5" (127mm) 3-jaw and a 7" (179mm) 4-jaw chuck, an 8" (200mm) face plate, steady and moving rests, single and 4-way tool posts, two dead centers, a splash guard and a tool box. An optional floor stand and a work light are available.

Figure 24. The Birmingham YCL-920BD lathe. Photo courtesy All Machine Tools.

29

Lathe sold in US	Swing	Centre distance	Headstock spindle bore	Motor	1 or 3 phase
Bolton CQ9332	12" (305mm)	24" (610mm)	1½" (38mm)	1hp (750W)	1
Grizzly G4002	12" (305mm)	24" (610mm)	1⁷/₁₆" (37mm)	1½hp (1.1kW)	1
Bolton CQ9332A	12" (305mm)	30" (762mm)	1½" (38mm)	1½hp (1.1kW)	1
Grizzly G4003	12" (305mm)	36" (914mm)	1⁷/₁₆" (37mm)	2hp (1.5kW)	1
Birmingham YCL-1236*	12" (305mm)	36" (914mm)	1½" (38mm)	2hp (1.5kW)	1
Grizzly G9249	12" (305mm)	37" (940mm)	1⁷/₁₆" (37mm)	2hp (1.5kW)	1
Harbor Fr 43681-2VGA	12" (305mm)	37" (940mm)	1½" (38mm)	2hp (1.5kW)	1
GML/MMX 1236 BG*	12" (305mm)	36" (914mm)	1½" (38mm)	2hp (1.5kW)	1
Birmingham YCL-1340*	13" (330mm)	40" (1,015mm)	1³/₈" (35mm)	2hp (1.5kW)	1
Harbor Fr 66164-1VGA*	13" (330mm)	40" (1,015mm)	1½" (38mm)	2hp (1.5kW)	1/3
Jet GHB-1340A*	13" (330mm)	40" (1,015mm)	1³/₈" (35mm)	2hp (1.5kW)	1
Jet GH-1340W	13" (330mm)	40" (1,015mm)	1½" (38mm)	3hp (2.2kW)	3
Bolton HA330*	13" (330mm)	40" (1,015mm)	1½" (38mm)	2hp (1.5kW)	1/3
Grizzly G9036*	13" (330mm)	40" (1,015mm)	1⁷/₁₆" (37mm)	2hp (1.5kW)	1

Table 2b. *The range of Far-East-built lathes sold in the US, in order of swing.*

Eight lathes have a 12" (305mm) swing. The **Bolton CQ9332** is complete with a 3-jaw chuck, a pair of MT dead centers and a set of tools. The **Bolton CQ9332A** has a greater distance between centers, a floor stand and the same accessories.

The **Grizzly G4002** has a 6" (150mm) 3-jaw and an 8" (200mm) 4-jaw chuck, a 10" (250mm) faceplate, a quick-change tool post with one tool holder, steady and moving rests, live and dead centers, a ½" (13mm) chuck and a toolbox. The **Grizzly G4003** has a longer distance between centers and the same accessories.

The **Birmingham YCL-1236-GH** is a gap-bed lathe and is fitted with a 6" (150mm) 3-jaw and an 8" (200mm) 4-jaw chuck, a 10" (250mm) faceplate, a 4-way tool post, steady and moving rests, a fixed center with sleeve, a splash guard and a tool kit.

The **Grizzly G9249** comes with a 6" (150mm) 3-jaw and an 8" (200mm) 4-jaw chuck, a 10" (250mm) faceplate, a 4-way and a rocker-style tool post, steady and moving rests, a fixed center with sleeve, a splash guard, a chip tray and a tool kit.

The **Harbor Freight 43681-2VGA** has 3- and 4-jaw chucks, a tool post, two dead and a live centre with sleeve, steady and moving rests and a toolbox with tools. There is an optional floor stand.

Three lathes, the **1236BG**, and the **1440BG** and **1440BGF** overleaf, can be purchased under both the **MMX** and **GMC** brands. However, they appear to be the same offerings.

The **GML/MMX 1236 BG** comes with a stand, a 6" (150mm) 3-jaw and an 8" (200mm) 4-jaw chuck, a 10" (250mm) faceplate, a 4-way tool post, steady and moving rests, two fixed centers with a sleeve, a splash guard and a tool box.

Next are twelve 13 x 40 lathes; some overleaf. The **Birmingham YCL-1340-GH** comes with a floor stand, 3- and 4-jaw chucks, an 11¾" (300mm) faceplate, a 4-way tool post, steady and moving rests, a fixed center with a sleeve, a splash guard and a tool box with tools.

Speeds		Dimensions length x width x height	Weight
60 - 1,600	12g	51³/₅" x 24³/₅" x 17¾" (1,311 x 625 x 451mm)	485lb (220kg)
70 - 1,400	9	53" x 26" x 23" (1,346 x 660 x 584mm)	913lb (415kg)
75 - 1,900	12g	59½" x 24³/₅" x 45" (1,511 x 625 x 1,143mm)	706lb (320kg)
70 - 1,400	9	61" x 23" x 23" (1,549 x 584 x 584mm)	917lb (417kg)
70 - 1,400	9	61" x 28" x 24" (1,549 x 711 x 610mm)	1,000lb (455kg)
15 - 1,200	12	64" x 27" x 49" (1,625 x 686 x 1,245mm)	1,136lb (516kg)
72 -1,600	18g	67" x 24¾" x 23¾" (1,702 x 629 x 605mm)	838lb (381kg)
0 - 1,400	9g	67" x 31" x 56" (1,702 x 787 x 1,422mm)	1,260lb (573kg)
70 - 2,000	8g	73¼" x 30" x 33" (1,861 x 762 x 838mm)	1,100lb (500kg)
60 -1,550	18g	67" x 19⁷/₈" x 24" (1,702 x 505 x 610mm)	1,058lb (481kg)
60 - 1,240	6g	71" x 32" x 45" (1,803 x 813 x 1,143mm)	954lb (434kg)
40 - 1,800	6g	74" x 30" x 48" (1,880 x 762 x 1,219mm)	2,080lb (945kg)
70 - 2,000	9g	78" x 31½" x 55" (1,981 x 800 x 1,397mm)	1,500lb (680kg)
70 - 2,000	8g	71½" x 30" x 53½" (1,816 x 762 x 1,359mm)	1,320lb (600kg)

* = a gap bed, g = a gearbox.

The **Harbor Freight 66164-1VGA** has 3- and 4-jaw chucks, a faceplate, a 4-way tool post, steady and travelling rests, MT live and dead centers, a drill chuck and a toolbox with tools.

The **Jet GHB-1340A** has constant headstock lubrication and comes with a 6" (150mm) 3-jaw and an 8" (200mm) 4-jaw chuck, a 12" (305mm) faceplate, a 4-way tool post, steady and follow rests, centers and a headstock center sleeve, a splash guard and a tool box with tools.

The **Jet GH-1340W** is available as a single or three-phase 230 volt machine and comes with a cabinet stand, 6" (150mm) 3-jaw and 8" (200mm) 4-jaw chucks, a 12" (305mm) faceplate, a 4-way tool post, steady and follow rests, centers and a headstock center sleeve, a work lamp, a splash guard, a removable chip tray and a tool box with tools.

The **Bolton HA330** has a 6" (150mm) 3-jaw chuck, two dead centers, a drill chuck with arbor and a set of tools. It comes with a coolant system and a stand.

The **Grizzly G9036** has constant headstock lubrication and comes with a 6" (150mm) 3-jaw and an 8" (200mm) 4-jaw chuck, a 12" (305mm) faceplate, a 4-way tool post, steady and follow rests, a splash guard, a chip pan and a stand.

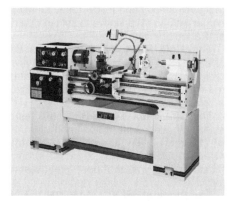

Figure 25. *The Jet GH-1340W lathe is a large machine. Photo courtesy JET/Walter Meier Manufacturing.*

Lathe sold in US	Swing	Centre distance	Headstock spindle bore	Motor	1 or 3 phase
Grizzly G9730/G9731*	13" (330mm)	40" (1,015mm)	1^3/$_8$" (35mm)	3hp (2.2kW)	1/3
Birmingham CT-1440G*	13" (330mm)	40" (1,015mm)	1½" (38mm)	2hp (1.5kW)	1
GML/MMX 1440BG*	13" (330mm)	40" (1,015mm)	1½" (38mm)	3hp (2.2kW)	1
GML/MMX 1440BGF*	13" (330mm)	40" (1,015mm)	1½" (38mm)	3hp (2.2kW)	1
Birmingham YCL1440*	13" (330mm)	40" (1,015mm)	1½" (38mm)	3hp (2.2kW)	3
Grizzly G4016*	14" (356mm)	40" (1,015mm)	1^7/$_{16}$" (37mm)	2hp (1.5kW)	1
Jet GH-1440W*	14" (356mm)	40" (1,015mm)	1½" (38mm)	3hp (2.2kW)	1/3
Harbor Fr 44995-2VGA*	14" (356mm)	40" (1,015mm)	1½" (38mm)	3hp (2.2kW)	3

Table 2c. *The range of Far-East-built lathes sold in the US, in order of swing.*

The **Grizzly G9730** and **Grizzly G9731** lathes only differ in that the former is designed to operate from a single-phase supply and the latter from a three-phase one. Both have built-in coolant systems and stands with splash guards. They come with 7" (175mm) 3-jaw and 8" (200mm) 4-jaw chucks, 10" (250mm) faceplates, extra 8" (200mm) D1-4 back plates, steady and follow rests, centers with sleeves, and tool boxes with a set of tools.

The **Birmingham CT-1440G** has a floor stand and comes with 3-jaw 160mm (6") and 4-jaw 200mm (8") chucks, a 4-way tool post, steady and follow rests, an MT3 center with an MT5 sleeve, a splash guard and a toolbox with tools.

The **GML/MMX 1440BG** has a 6" (150mm) 3-jaw and an 8" (200mm) 4-jaw chuck as well as a 12" (300mm) face plate, a 4-way tool post, steady and follow rests, a splash guard and a stand.

The **GML/MMX 1440BGF** has a floor stand with work light, a coolant system, a 6" (150mm) 3-jaw and an 8" (200mm) 4-jaw chuck as well as a 12" (300mm) faceplate, a 4-way tool post, steady and follow rests and a splash guard.

The **Birmingham YCL-1440-GH** has a coolant system and a floor stand with a work light, a 6" (150mm) 3-jaw and an 8" (200mm) 4-jaw chuck as well as a 12" (300mm) faceplate, a 4-way tool post, steady and follow rests, centers and a center sleeve, a tool box with a set of tools, a splash guard and removable chip tray.

The final three 14 x 40 lathes vary but all have a floor stand, Camlok 6" (150mm) 3-jaw and 8" (200mm) 4-jaw chucks.

The **Grizzly G4016** has a 4-way tool post, steady and follow rests, a live and two MT3 dead centers, a splash guard and a chip tray.

The **Jet GH-1440W-1** is available as a single or three-phase 230volt machine. It has a coolant system and a work light, a 12" (300mm) face plate, a 4-way tool post, steady and follow rests, centers and a center sleeve, a removable chip tray and a splash guard.

The **Harbor Freight 44995-2VGA** has a coolant system, a 10" (250mm) face plate, steady and follow rests, a 4-way tool post, a dead center and a reduction sleeve, a splash guard, a chip pan and a work light.

Second-hand lathes

The use of CNC lathes by industry and the closure of many engineering-based training schemes have released manual machines onto the second-hand market. The quality of many of them is such that

Speeds		Dimensions length x width x height	Weight	
105 - 2,000	8g	70" x 32½" x 50" (1,778 x 825 x 1,270mm)	1,475lb	(670kg)
70 - 1,350	12g	73¼" x 30" x 30" (1,861 x 762 x 838mm)	800lb	(364kg)
70 - 2,000	8g	68" x 32" x 57" (1,727 x 813 x 1,448mm)	1,360lb	(618kg)
70 - 2,000	8g	70" x 35" x 59" (1,778 x 889 x 1,499mm)	1,730lb	(786kg)
45 - 1,800	16g	72" x 31" x 49" (1,829 x 787 x 1,245mm)	2,000lb	(909kg)
78 - 2,100	8g	71½" x 26" x 52½" (1,816 x 660 x 1,334mm)	1,261lb	(573kg)
40 - 1,800	6g	73¾" x 29½" x 47¼" (1,873 x 750 x 1,200mm)	2,191lb	(996kg)
45 - 1,800	16g	75½" x 29" x 47" (1,920 x 740 x 1,194mm)	1,650lb	(750kg)

* = a gap bed, g = a gearbox.

they will last a model engineer's lifetime. Also, most home workshops are cleared on the death of the incumbent providing a flow of usually well-maintained machines.

While many of these lathes are no longer in production, a number of current machines are also offered second-hand. Providing spares for out-of-production lathes has proved good business for the original manufacturers and companies that specialise in supporting such lathes.

Sources of second-hand lathes, and parts for them, are the companies that focus on their sale, advertisements in model engineering magazines and Ebay. All of them offer a choice of lathes in a range of sizes from various manufacturers.

Boxford

Second-hand examples of the **Boxford Model A, VSL, ME10, CSB, Model T** and **Model AUD, BUD, CUD** and **TUD, STS** and **1130** lathes are still obtainable and are popular with model engineers. And Boxford continues to supply spares for their **AUD, BUD, CUD** and **ME10** lathes.

Colchester and Harrison

The **Colchester Student** and **Master** lathes, the latter with longer beds, and the **Harrison M250** all suit the larger home workshop. Unfortunately, neither company

Figure 26. *A typical example of a second-hand Boxford lathe; the CUD 111. Photo courtesy Home and Workshop Machinery.*

Figure 27. *A nice example of a Colchester Student 1800. Photo courtesy Home and Workshop Machinery.*

Figure 28. *A Hardinge HLV toolroom lathe. Photo courtesy Rondean Machinery.*

offers spares for these machines but they can be obtained from several specialist spares suppliers.

Emco

Early models of the **Unimat** lathe are still popular; perhaps more so than larger lathes like the **Emcomat 7** and **8**, the range of **Prazimats**, **Maximats** and the **Compact 10**. And Emco still provide spares and manuals for the **Unimat SL** and **Mk 3**, **Emcomat 7**, **V8**, **8.4** and **8.6**, **Maximat V10P**, **Super 11** and **V13**.

Hardinge/Bridgeport

The **Hardinge HLV** tool room lathe shown in Figure 28 has only recently ceased to be produced after 60 years. Second-hand versions of this machine are desirable and continue to be supported by Bridgeport.

Myford

Myford still produce spares and documents for their **ML7** (not clutch parts), **ML7R**, **ML10**, **Speed 10**, **Diamond 10** and **254 Plus** lathes; the earliest over 60 years old.

Figure 29. *A very early Myford ML1 lathe dating back to the 1930s.*

South Bend

Lathes like the **South Bend Light Ten** and **Heavy 10"**, and later examples of the **9" Workshop** continue to be supported by South Bend. Older models have to rely on parts from spares-supply companies, with many of these located in the United States.

Far East lathes

Warco has been importing several of their Asian lathes into the UK for the last twenty-five years. Some of their lathes, as well as those of other importers from the Far-East, are now appearing on the second-hand market. Providing that they are in reasonable condition and spares are still available, they can prove to be very economic solutions.

Figure 30. *A 1952 South Bend 9" Model A lathe. Photo courtesy lathes.co.uk.*

Chapter 2

Combination lathes and mills

The acquisition of a combined lathe and mill or a multi-function machine (basically an innovation from the Far East) offers a number of attractions when compared with a separate lathe and mill. The advantages include a significant saving of space and this almost inevitably is combined with a cost reduction when contrasted with the price of two machines.

The biggest drawback is likely to be that, when using the milling function, there is a limit to the size of component that can be fitted on the cross slide and machined. Also, the movement of the top slide across the lathe bed is liable to be limited. More (or less) setting up time may be needed when changing from turning to milling operations on a work piece and vice-versa.

Three UK and three US importers offer a choice respectively of eight and seven different combination lathe and milling machines. All of these machines employ a comparable design philosophy.

A similar solution to the same problem is to fit a separate milling attachment to an existing lathe. This topic is covered later in this chapter and offers a possible solution for those who already own a lathe.

Not every lathe can easily have a milling head attached but at least six companies offer this solution for a number of their

lathes. Four imported lathes of Far East origin can be fitted with an attachment as also can six lathes that are manufactured in Europe.

In addition, three machines, the **Chester Cestrian**, the **Golmatic MD23UWG** and **MD24**, offer a completely different type of solution. They are genuine multi-function machines that can be configured to give the facilities of a lathe, a vertical or a horizontal mill.

All of the machines in this chapter use single-phase mains electricity.

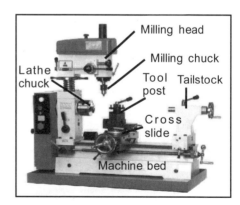

Figure 1. *The main parts of a combined lathe/ mill. Photo courtesy Chester.*

Details	Chester Model B Super Multifunction	Chester Centurion 3-in-1	Chester Centurion 3-in-1 Long Bed
Lathe centre height	210mm (8¼")	210mm (8¼")	210mm (8¼")
Distance between centres	520mm (20½")	500mm (19⅝")	800mm (31½")
Lathe speeds	160 - 1,360rpm	160 - 1,360rpm	160 - 1,360rpm
Number of lathe speeds	7	7	7
Milling speeds	117 - 1,300rpm	120 - 3,000rpm	120 - 3,000rpm
Number of milling speeds	14	16	16
Milling table width	300mm (12")	475mm (18¾")	475mm (18¾")
Milling table depth	150mm (6")	160mm (6¼")	160mm (6¼")
Cross-slide travel	185mm (7¼")	200mm (8")	200mm (8")
Top-slide travel	80mm (3⅛")	80mm (3⅛")	80mm (3⅛")
Throat	135mm (5⅓")	306mm (12")	306mm (12")
Spindle stroke	80mm (3⅛")	110mm (4⅓")	110mm (4⅛")
Motors	1 x 550W (¾hp)	2 x 550W (¾hp)	2 x 550W (¾hp)
Length	1,090mm (43")	1,570mm (61¾")	1,570mm (61¾")
Width	690mm (27⅛")	480mm (18⅞")	480mm (18⅞")
Height	930mm (36⅔")	955mm (37⅔")	955mm (37⅔")
Weight	155kg (341lb)	250kg (550lb)	300kg (660lb)

Table 1. *The range of Far-East-manufactured combination lathe/mills available in the UK.*

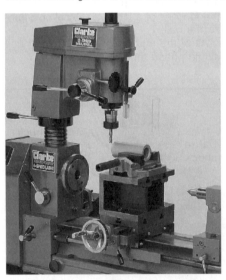

Figure 2.*The Clarke CL500M Multi-function machine. Photo courtesy Chronos.*

Multi-function machines

All the Chester multi-function models in Table 1 have the same centre height but varying centre distances. The **Chester Model B Multifunction** has the smallest milling table and throat. It has a 100mm (4") 3-jaw chuck, two dead centres, a drill chuck, a dual-purpose compound slide/ machine vice, a 4-way indexing tool post and a set of lathe tools.

The **Chester Centurion 3-in-1 Long Bed** has a larger centre distance than the **Chester Centurion 3-in-1**. Both employ separate motors for turning and milling and have a power cross feed. They also come with a similar set of items to the **Model B** but with a 125mm (5") 3-jaw chuck.

Options for all three machines include a 125mm (5") 4-jaw chuck, a 200mm (8") faceplate, a vertical slide, a quick-change tool post, a coolant system and a stand.

Clarke CL500M	Warco WMT 300/1	Warco WMT 300/2	Warco WMT 500	Warco WMT 800
152mm (6")	150mm (6")	150mm (6")	150mm (6")	210mm (8¼")
430mm (17")	500mm (19⅝")	500mm (19⅝")	500mm (19⅝")	800mm (31½")
170-1,630rpm	160 - 1,600rpm	160 - 1,600rpm	160 - 1,600rpm	160 - 1,360rpm
6	6	6	6	6
130-1,660 rpm	125 - 1,600rpm	125 - 1,600rpm	360 - 2,150rpm	120 - 3,000rpm
12	12	12	12	16
200mm (8")	204mm (8")	430mm (17")	430mm (17")	475mm (18¾")
150mm (6")	150mm (6")	150mm (6")	150mm (6")	160mm (6¼")
–	150mm (6")	240mm (9½")	240mm (9½")	200mm (8")
–	75mm (3")	75mm (3")	75mm (3")	75mm (3")
320mm (12½")	171mm (6¾")	171mm (6¾")	171mm (6¾")	171mm (6¾")
92mm (3²/₃")	92mm (3⅝")	92mm (3⅝")	92mm (3⅝")	110mm (4⅛")
1 x 550W (¾hp)	1 x 550W (¾hp)	1 x 550W (¾hp)	2 x 550W (¾hp)	2 x 550W (¾hp)
1,100mm (43⅓")	1,070mm (42")	1,070mm (42")	1,067mm (42")	1,570mm (61¾")
600mm (23²/₃")	580mm (22¾")	580mm (22¾")	584mm (23")	610mm (24")
880mm (34²/₃")	870mm (34¼")	870mm (34¼")	787mm (31")	980mm (38½")
164kg (361lb)	150kg (330lb)	165kg (363lb)	220kg (484lb)	294kg (647lb)

The **Clarke CL500M** is the smallest machine in terms of centre height and centre distance but not in terms of its dimensions or its weight. It is fitted with a 3-jaw chuck and a 4-way tool post. Options include a 4-jaw chuck, a milling chuck, a faceplate, a machine block, fixed and moving steadies, a revolving centre, a tailstock chuck and a floor stand.

The **Warco WMT 300/1** and **300/2** are identical apart from the width and amount of movement of their milling tables. The **Warco WMT 500** and **WMT 800** are both based on the **WMT 300** but feature a power cross feed and a separate milling motor; the **WMT 800** also having a longer lathe bed. All four come with a 125mm (5") 3-jaw chuck, a 200mm (8") faceplate, fixed and travelling steadies, two dead centres, a drill chuck, a vice, a face cutter with replaceable tips and a set of tungsten-carbide-tipped lathe tools. Options include a 150mm (6") 4-jaw chuck, a riser block, a quick-change tool post, a live centre, a tailstock die holder, a floor stand and an inverter drive to give variable speeds from 16 - 1,920rpm in 6 overlapping steps.

Figure 3.*The Warco WMT 300/1 Multi-function machine. Photo courtesy Warco.*

Details	Bolton AT125	Harbor Freight 5980-0VGA	Bolton AT520
Lathe swing	4⁹/₁₀" (124mm)	7¼" (184mm)	12" (305mm)
Distance between centers	7" (178mm)	14" (356mm)	20" (508mm)
Bed width	–	–	–
Headstock spindle bore	⅓" (8mm)	¾" (20mm)	1" (26mm)
Lathe speeds	560 - 2,500rpm	430 - 2,000rpm	160 - 1,600rpm
Number of lathe speeds	5	5	6
Milling speeds	500 - 2,500rpm	430 - 1,500rpm	315 - 2,000rpm
Number of milling speeds	10	4	9
Milling table width	3⅓" (85mm)	7⅞" (200mm)	3⅓" (89mm)
Milling table depth	2⅓" (59mm)	6⅛" (155mm)	6⅛" 155mm
X-axis travel	–	16" (406mm)	12¹/₁₆" (306mm)
Y-axis travel	–	5¼" (133mm)	8½" (216mm)
Throat	7" (178mm)	–	–
Spindle stroke	–	3¼" (83mm)	3¼" (83mm)
Lathe motor	⅓hp (250W)	½hp (370W)	¾hp (550W)
Milling motor	–	½ hp (370W)	¾hp (550W)
Length	20½" (520mm)	35⅝" (890mm)	42" (1,067mm)
Width	11¾" (300mm)	23¼" (580mm)	20" (508mm)
Height	16⅛" (410mm)	31⅝" (810mm)	29" (737mm)
Weight	48½lb (22kg)	326lb (148kg)	441lb (200kg)

Table 2. *The range of Far-East-manufactured combination lathe/mills available in the US.*

Far East machines in the US

Table 2 lists seven combo lathe/mills that are imported into the US from the Far East by three different suppliers. Many of these machines are similar those imported into the UK although they are suited to US mains voltages and frequencies. Once again, imperial measurements are quoted before their metric equivalents.

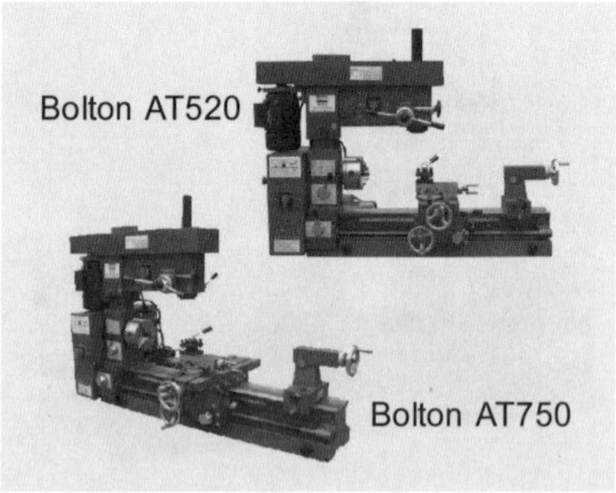

The **Bolton AT125** is the smallest multi-function machine in every respect. It is supplied with a 3-jaw chuck, a drill chuck and two dead centers.

The **Harbor Freight 5980-0VGA** is a twin-motor machine and comes with a drill chuck and two dead centers.

The **Bolton AT750** is a long-bed version of the **Bolton AT520**. Both have a pair of motors, power cross feed and are fitted with a 3-jaw chuck, a drill chuck and two dead centers. Options for both machines include a 4-jaw chuck and a faceplate, a live center, steady and follow rests, a riser block, a collet chuck and a machine vise.

Bolton AT520

Bolton AT750

Figure 4. *The Bolton AT520 and AT750 multi-function machines. Photo courtesy Bolton.*

Bolton AT750	Grizzly G4791 Large Combo Lathe	Grizzly G4015Z Combo Lathe/Mill	Grizzly G9729 Combo Lathe/Mill
12" (305mm)	12" (305mm)	16½" (419mm)	16½" (419mm)
29½" (749mm)	39" (991mm)	19$^{1}/_{5}$" (488mm)	31" (787mm)
–	6$^{1}/_{8}$" (156mm)	5$^{1}/_{8}$" (130mm)	5$^{1}/_{8}$" (130mm)
1" (26mm)	1½" (38mm)	¾" (19mm)	1$^{1}/_{8}$" (29mm)
160 -1,600rpm	345 - 1,960rpm	185 - 1,455rpm	175 - 1,425rpm
6	9	7	7
315- 2,000rpm	435 - 2,345rpm	135 - 1,370rpm	135 - 1,500rpm
9	9	14	16
5$^{7}/_{8}$" (149mm)	7$^{7}/_{8}$" (199mm)	7¼" (184mm)	6¼" (159mm)
16¾" (425mm)	16$^{7}/_{8}$" (427mm)	6" (152mm)	18¾" (476mm)
12$^{1}/_{16}$" (306mm)	35" (889mm)	13" (330mm)	29" (737mm)
8½" (216mm)	9" (227mm)	4½" (114mm)	3¾" (95mm)
–	14¼" (362mm)	10" (254mm)	12" (306mm)
3¼" (83mm)	5" (127mm)	3½" (89mm)	4$^{5}/_{16}$" (110mm)
¾hp (550W)	¾hp (550W)	¾hp (550W)	¾hp (550W)
¾hp (550W)	1½hp (1.1kW)	–	¾hp (550W)
53" (1,346mm)	69" (1,753mm)	42" (1,067mm)	58" (1,473mm)
20" (508mm)	25" (635mm)	23" (584mm)	40½" (1,029mm)
28$^{1}/_{3}$" (720mm)	69" (1,753mm)	35" (889mm)	40" (1,016mm)
>665lb (302kg)	1,110lb (500kg)	440lb (200kg)	525lb (239kg)

The **Grizzly G4791 Large Combo Lathe** has the largest distance between centers and is nearly twice the weight of any other of these combo machines. It comes with a 6" (150mm) 3-jaw chuck, an 8" (200mm) 4-jaw chuck and a faceplate, two dead centers, a steady and follow rests, a 4-way tool post, a ½" (13mm) drill chuck, a face-mill cutter and a floor stand. Gear changes are required for imperial thread cutting.

The **Grizzly G4015Z Combo Lathe/Mill** uses a single motor for both turning and milling. Standard equipment includes a 4" (100mm) 3-jaw chuck, a 9½" (229mm) faceplate, ½" (13mm) drill chuck, two dead centers, a built-in vise and a 4-way tool post.

The **Grizzly G9729 Combo Lathe/Mill** features a 4-way turret tool post, steady and follow rests, a display of spindle speed and power cross feed. Standard accessories include a 5" (125mm) 3-jaw chuck, a ½" (13mm) drill chuck, two dead centers and a 9" (229mm) faceplate.

Figure 5. The Grizzly G9729 combo lathe/mill. Photo courtesy Grizzly Industrial Inc.

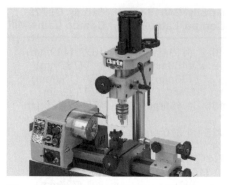

Figure 6. *The Clarke CL251MH milling head fits very neatly onto a CL250M lathe. Photo courtesy Chronos.*

Milling attachments

Figure 7. *The Emco milling attachment fitted to a Unimat 4 lathe. Photo courtesy Pro Machine Tools Ltd.*

Adding a milling facility to an existing lathe, instead of buying a milling machine, gives several advantages. The main benefit is the lack of need for further space beyond that already occupied by the lathe, and the cost will be lower as a compound table is not needed. The top-slide size and travel will be those of the lathe; usually smaller than those of a milling table. There are also implications when changing from lathe work to milling.

Axminster
The Axminster Sieg C4 and **C6 Milling attachments** for the **Sieg C4** and **C6 lathes** are based on the **Axminster X1 Micro Mill** and **X2 Milling Machines** (see page 54). The **X1** has a 150W ($^1/_5$hp) motor providing speeds of 0 - 2,000rpm and adds 17kg (37lb) to the lathe weight, The **X2** has a 350W ($^1/_2$hp) motor with spindle speeds of up to 2,500rpm.

Chester
For users of the **Chester Cobra lathe**, the **Cobra mill** (see page 54) also comes as a milling attachment. It has a 150W ($^1/_5$hp)

motor giving speeds from 100 - 2,000rpm. Its vertical travel is 225mm (9") and it increases lathe weight by 30kg (66lb).

Clarke
The **Clarke CL251MH Mill Head** has a 150W ($^1/_5$hp) mains motor that gives speeds from 10 - 1,300rpm. The maximum distance from the chuck to the **CL250M lathe** is 168mm ($6^5/_8$"), the distance from the column to the quill is 105mm ($4^1/_8$") and the quill travel is 30mm ($1^1/_8$"). The attachment increases the lathe height to 575mm ($22^2/_3$") and weighs 18kg (40lb).

Emco
The milling head from the **Unimat Milling Machine**, described on page 48, has a 35mm ($1^1/_3$") column and can be fitted to the **Unimat 4** or **Compact 5 lathes**. Its 95W ($^1/_8$hp) mains motor provides six speeds from 280 - 2,500rpm. The head tilts up to 90° each way. The maximum distance from the table to the centre of the milling spindle is 310mm (12¼"), the quill stroke is 60mm ($2^1/_3$") and the attachment weighs 6.5kg ($14^1/_3$lb).

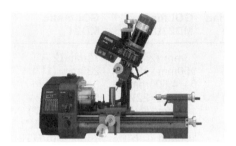

Figure 8. *A Proxxon PF 400 mill on a PD 400 lathe. Photo courtesy BriMarc/Proxxon.*

Figure 9. *A Wabeco DF1680 mill on a D6000 E lathe. Photo courtesy Pro Machine Tools Ltd.*

Proxxon

The **PF 230 Milling Drilling Head** and the **PF 400 Mill/Drill Head** will fit on the **PD 230/E** and **PD 400 lathes** that already have mounting flanges. They have the same specifications to the **FF 230** and **FF 400 Micro Millers** described on page 49. A table measuring 110 x 70mm (4$^{1}/_{3}$" x 2¾") replaces the top slide on the **PD 230/E** and one measuring 150 x 110mm (6" x 4$^{1}/_{3}$") on the **PD 400**.

Wabeco

The **Wabeco DF1680 E Universal Drill/ Mill Unit** will fit the **Wabeco D6000 E** and **D6000 E high speed lathes**. Longitudinal and transverse movements are 500 x 140mm (19$^{2}/_{3}$" x 5½") while the vertical range is 280mm (11"). A 1.4kW (2hp) milling motor gives infinitely variable speeds from 140 to 3,000rpm. The head swivels 90° either way. The cross-slide table measures 270 x 150mm (9$^{7}/_{8}$" x 6") and the drilling stroke is 55mm (2"). Maximum distance between the table and head is 280mm (11"). The milling attachment weighs 49kg (108lb).

Other multi-function machines

Some types of multi-function machine are more than just lathes with vertical mills.

Chester

The **Chester Cestrian Multi-function machine** can be configured as a lathe or for vertical or horizontal milling. As a vertical mill, it can tilt its head from 0 - 95° to mill at an angle. It has a geared head providing variable spindle speeds. The milling capacity is 40mm (1½") in vertical mode and 70mm (2¾") when horizontal. A detailed specification is given in Table 3 overleaf.

It comes complete with a drawbar, drill chuck and arbor, a horizontal arbor, table clamps and a toolbox.

Figure 10. *The Chester Cestrian Multi-function machine set up for turning. Photo courtesy Chester.*

41

Details	Chester Cestrian Multi-function	GOLmatic MD23UWG	GOLmatic MD24
Lathe centre height	75mm (8¼")	73mm (2⁷⁄₈")	73mm (2⁷⁄₈")
Distance between centres	260mm (10¼")	350mm (13¾")	280mm (11")
Speeds	0 - 2,800rpm	0 - 4,500rpm	0 - 4,000rpm
Table width	470mm (18½")	570mm (22½")	570mm (22½")
Table depth	160mm (6¼")	160mm (6¼")	160mm (6¼")
Sideways travel	350mm (13¾")	500mm (19²⁄₃")	350mm (13¾")
Fore and aft travel	160mm (6¹⁄₃")	160mm (6¹⁄₃")	160mm (6¹⁄₃")
Z-axis, vertical	330mm (13")	330mm (13¾")	400mm (15¾")
Z-axis, horizontal	-	600mm (23²⁄₃")	600mm (23²⁄₃")
Quill stroke	60mm (2¹⁄₃")	60mm (2¹⁄₃")	60mm (2¹⁄₃")
Spindle rotation	95°	90°	90°
Motor	550W (¾hp)	1.1kW (1½hp)	1.1kW (1½hp)
Length	1,200mm (47¼")	480mm (18⁷⁄₈")	480mm (18⁷⁄₈")
Width	815mm (32")	400mm (15¾")	400mm (15¾")
Height	1,120mm (44")	1,400mm (55")	1,400mm (55")
Weight	300kg (660b)	300kg (660b)	320kg (660b)

Table 3. Three truly multi-function machines available in the UK.

GOLmatic

The **GOLmatic MD23UWG** will suit quality-orientated model engineers. It is a universal desktop machine that can be configured to turn, drill, mill vertically or horizontally, engrave, broach or grind.

Options include central lubrication, power drives for all three axes, a 100mm (4") 3-jaw lathe chuck, a quick-action 4-way tool post, a revolving centre, a conversion kit for horizontal milling, a broaching head and a 3-axis digital-readout display.

The **GOLmatic MD24** is also a universal machine, developed from the **MD23** but with a larger column and base. It has a longer Z axis and a central lubrication system. It is also available with an extra-long table of 720mm (28¹⁄₃") that moves 600mm (23²⁄₃"). It has digital displays for table and quill movement. Its footprint is 1,200mm x 1,000mm (39¹⁄₃" x 47¼").

It quickly converts from one function to another and comes in a protective cabinet. It has a tailstock, a 4-way tool post, an

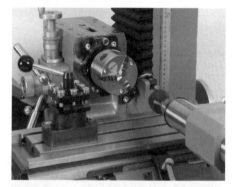

Figure 11. The GOLmatic MD23UGW in use as a lathe. The MD24 is very similar. Photo courtesy Pro Machine Tools Ltd.

80mm (3¹⁄₈") 3-jaw chuck, a fixed centre, a three-part vice system, an outrigger for horizontal milling, an acrylic splash guard, a set of clamping screws and a set of tools. Options include automatic X and Y feeds, a three-ratio gearbox and an LCD speed display.

Chapter 3

Milling machines

The main difference between milling and turning is that a mill's cutting tool revolves against a work piece whilst a lathe's work piece rotates against a tool. It is only in recent years that milling machines, and particularly vertical milling machines, have become affordable and widely available for home use. So, while some milling can be carried out in a lathe or on the cross slide of a combination lathe/mill, a dedicated milling facility has much to recommend it, especially if large areas need to be milled.

A milling machine can shape metal into flat, curved or irregular shapes, cut dovetails and slots with a range of profiles as well as drill holes. A motor-driven spindle holds a cutter and beneath is the work piece fixed to a table. Milling involves moving the work piece or cutter in three dimensions; left/right, forwards/backwards and up/down, producing a path along which metal is removed. The movements are precisely controlled with slides and lead screws either by hand or with power. The spindle may be fixed in position and the table moved in three dimensions or the spindle may be able to move vertically while the table can move horizontally in two dimensions.

While most American and European lathe manufacturers also make milling machines, their range of products is far more limited than those of Asian machine-tool suppliers. This may be because model engineers are renowned for their limited budgets and space restrictions or because some milling can be done in the lathe. Thus bench milling machines are popular because of their relatively small footprint.

Figure 1. *A classic bench-mounted vertical mill, the Myford VM-B.*

43

Figure 2. *The original Bridgeport knee-type, vertical milling machine. Photo courtesy lathes.co.uk.*

In the UK this is often the critical consideration. In the US, there is usually enough room for larger machines but, for practical space and portability reason, only machines weighing less than 1,000kg (2,200lb) are included in this book.

While Myford, many years ago, set the standard by which small lathes were judged, Bridgeport set similar standards for vertical-milling machines. The well-loved Bridgeport vertical-knee mill that was introduced in the first half of the twentieth century became the brand leader for such machines. Its revolving turret and ram allow the milling head to move over the table while the table itself can be vertically moved up and down. This idea has been copied by most other manufacturers.

A vertical mill has its spindle mounted parallel to its column. Most vertical-milling machines have a sliding head that can be moved up and down and may also allow

the head to tilt sideways, on some up to 90° each way. Thus angular surfaces can be milled and some limited horizontal milling may be possible. A horizontal attachment can also be clamped to the quill with bevel gears to turn the drive through 90°.

Milling machines are classed as bed mills if the table only moves horizontally; the head and spindle moving up and down. They are turret mills if the spindle remains stationary in all axes during cutting while the table moves in three dimensions; parallel and perpendicular to the spindle axis.

Knee-type milling machines are larger and heavier than bench-mounted ones and are fitted with a vertically adjustable table on a saddle supported by a knee. The knee and table may be clamped in any desired position. The saddle moves in and out and the table traverses right and left on the saddle to position the work. The table may be manually controlled or power fed. Ram-type milling machines have their spindle mounted on a movable housing on top of the column to allow the milling cutter move forward or backwards.

A horizontal mill is a different type of machine and has its cutter mounted on an arbor with its axis parallel to the table. Its column incorporates a motor and gearing, with an adjustable overhead arm that projects forward from the top of the column to support the arbor that is mounted on the spindle. Simplex mills have just one spindle while duplex mills have two. Many of these mills also feature a tilting table that allows milling at shallow angles. Their real advantage lies in their arbor-mounted cutters (see page 99). Because these cutters are well support by the arbor, quite heavy cuts can be taken, enabling rapid material removal (more of a benefit in mass production than for home working). Several cutters may be ganged together on an arbor to mill complex shapes. Special cutters can produce grooves, bevels or any

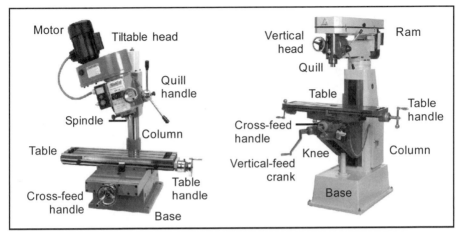

Figure 3. *The various parts of bench-mounted, (left) and knee-type, (right) milling machines. Photos courtesy Warco (left) and Chester (right).*

section desired but tend to be expensive. It is also relatively easy to cut gears on a horizontal mill.

The classic Bridgeport layout offers the following advantages when compared with horizontal milling machines:

1. It can advance and retract the cutter without having to raise and lower the milling table allowing rapid tool withdrawal to clear chips or check progress.

2. Tactile feedback from the quill feed indicates how the tool is cutting so that the feed can be optimized and the risk of tool breakage reduced.

3. The ability to cut angles by tilting the head rather than mounting the work at an angle on the table.

4. The one-piece head eliminates the need for complex gearing.

5. Reduced cost by using lower-power motors and smaller, less-rigid castings as well as smaller tools.

It is, therefore, unsurprising to find that the horizontal mill has all but disappeared from the catalogues of the companies supplying model engineers, apart from second-hand machines and a few new combined vertical/horizontal machines.

The considerations already listed in the introduction should help to define the budget, the size of mill, its quality, whether to choose a metric or imperial machine and any preference for a new or second-hand machine.

The main factors when selecting a milling machine are whether to go for a vertical, horizontal or combined mill, and if a vertical one, whether to choose a bench-mounted, a free-standing knee-type or a turret machine. Other factors include the table size and its horizontal movement, the vertical travel of the milling head or table and the quill movement; also the location of the vertical hand wheel. A power quill-feed tied to spindle rotation speed and table power feeds can be advantages. The ability of the head to be tilted up to 45° or even 90° each side of the vertical increases the flexibility of a machine. The power of the drive motor and how the spindle speed is varied (with belts on pulleys, a gearbox or electronic-speed control) are issues to

45

Figure 4. *The Bridgeport Series 1 knee-type mill. Photo courtesy Rondean Machinery.*

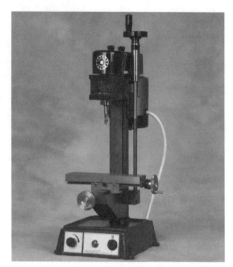

Figure 5. *Cowells vertical-milling machine. Photo courtesy Cowells Small Machine Tools Ltd.*

consider. Finally the number of table slots, their sizes and the spindle-hole size and taper are often crucial if suitable accessories are already owned. Quality machines will come with a collet chuck while lower-cost machines may only have a Jacobs chuck as standard.

Vertical-milling machines

The widespread use of vertical-milling machines in home workshops is quite a recent phenomenon and is, in large part, due to their versatility. They can take end and slot mills, face-milling cutters and any other form of cutter that can be mounted on a suitable mandrel. There are many manufacturers of milling machines located both in the industrialised nations and in the Far East. Typical examples are Myford in the UK, Bridgeport in the US, Emco in Austria, Wabeco in Germany and many Chinese mills badged by importers both across Europe and in North America.

Bridgeport

Hardinge Inc, an American company, manufactures a range of milling machines including the Series 1 Bridgeport machine that set the standard by which other turret vertical-milling machines are judged. Over one third of a million Series 1 machines have been produced over the last 60 years. It employs a scraped grey cast-iron construction and is powered by a 3hp (2.2kW) three-phase motor giving a speed range from 60 - 4,200rpm, that is infinitely variable above 500rpm. The table is 48" x 9" (1,219mm x 299mm) and can move 36" x 12" (914mm x 305mm) horizontally and 16" (406mm) up and down. In addition, the quill can move 5" (127mm) up and down and 12" (305mm) back and forth.

The mill occupies 98" x 63" (2,500mm x 1,600mm) floor space, is 80" (2,100mm) high and weighs 1,950lb (885kg). It has a

46

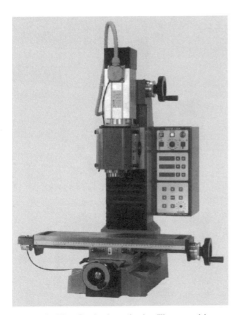

Figure 6. *The Ceriani vertical-milling machine with digital display and power feeds. Photo courtesy Chester.*

Figure 7. *An Emco F1 vertical-milling machine. Photo courtesy Emco.*

300mm (11⁷/₈"). The machine measures 780 x 850 x 1,000mm (30¾" x 33½" x 39¹/₃") and weighs 150kg (330lb). It comes with a drill chuck and draw bar.

Cowells
A small but accurate British machine, the **Cowells Vertical Milling Machine** is driven by a single-phase 125W (¹/₆hp) motor, giving electronically variable speeds from 40 - 4,000rpm, and can be supplied for European or US mains. The table size is 203mm x 50mm (8" x 2") and it moves 150mm x 80mm (6" x 3¹/₈"). Vertical head travel is 164mm (6½"). The machine measures 260mm x 343mm (10¼" x 13½"), is 420mm (16⁵/₈") high and weighs 25Kg (55lbs).

Emco Hobby Machines
The **Emco Maximat F1-P Co-ordinate Milling and Drilling Machine** is Austrian and is driven by either a 180W (¼hp) single- or 250W (¹/₃hp) three-phase motor giving four speeds from 350 - 1,800rpm, altered by belt changing. The maximum distance from the spindle nose to the table is 370mm (14½"). The milling head can be swivelled through 360°. The table measures 630mm x 150mm (25" x 6") and

one-shot lubrication system and uses chrome-plated ways and gib strips. The options include a two-axis digital readout, a power drawbar for collets, a quick-change spindle, a work light, a coolant system, splash backs and a chip pan.

Ceriani
The Ceriani David variable-speed milling machine is made in Italy and the basic mill has manual feeds. Options include a digital display, power feeds for all three axes and a floor stand. A single-phase 1.5kW (2hp) motor provides variable spindle speeds from 100 - 3,000rpm. The head can tilt 90°, the table measures 580 x 150mm (22⁷/₈" x 6") and it can move 420 x 160mm (16½" x 6¼"). Vertical travel is 300mm (11⁷/₈") and the maximum distance from the spindle to the table is also

Figure 8. *The Myford VM-E vertical mill being use to repair full-size engine parts.*

Figure 9. *The Proxxon FF 230 vertical-milling machine. Photo courtesy BriMarc/Proxxon.*

has three T-slots. It can travel 380mm x 140mm (15" x 5½") and the head moves 370mm (14½") up and down. The quill stroke is 50mm (2"). The mill measures 700mm x 800mm (27½" x 31½") and is 1,100mm (43^1/$_3$") high. It weighs about 120kg (216lb). Accessories include a machine vice, an angle plate, a rotary table, collets, a drill chuck and a fly cutter.

The **Unimat Milling Machine** has a 65/ 95W (^{1}/$_8$hp) motor giving six speeds from 280 - 2,500rpm. It has a 35mm (1^1/$_3$") column and the head will tilt up to 90° either way. The maximum distance from the table to the milling spindle is 200mm (7^7/$_8$") and the quill stroke is 60mm (2^1/$_3$"). The co-ordinate table is 280mm x 95mm (11" x 3¾") with a travel of 190mm x 100mm (7½" x 4"). It weighs 22kg (48lb).

Myford

The **Myford VM-E** is the solitary milling machine produced by Myford and comes

in both metric and imperial variants. It is powered by a 750W (1hp) motor that can be either for single- or three-phase mains AC supply. It has nine spindle speeds from 180 to 2,860rpm, achieved with belt changing. The motor is on a hinged platform to ease speed changing. The optional Vari-Speed controller gives four variable speed ranges; 40 - 300rpm, 70 - 524rpm, 228 - 1,714rpm and 370 - 3,000rpm.

The milling head swivels 45° either way about both horizontal and vertical axes. The machine has a substantial column with full-length V-slideways. With a cast-iron base, it is mounted on a sheet-steel stand incorporating a chip tray.

The surface-ground table is 760mm x 180mm (30" x 7^1/$_8$") with three T-slots. An optional table power-feed can be supplied with new machines or user retro-fitted to existing machines. The table can move 495mm x 190mm (19½" x 7½") and 310mm (12¹/₅") up and down. To special order, machines can be factory fitted to increase the maximum height from table to spindle nose by 100mm (4"). A two- or three-axis digital readout is available as an optional extra.

The wide range of accessories includes a dividing head, a horizontal and vertical rotary table, a machine vice, a collet nose piece and collets, a Posilock chuck, a boring head, a fly cutter, a halogen work light and coolant equipment.

Proxxon

The **Micro Miller MF 70** is an extremely small German machine fitted with a 100W (¹/₈hp) mains motor giving continuously variable spindle speeds of 5,000 - 20,000rpm. Six collets from 1mm - 3.2mm (³/₆₄" - ¹/₈") are included. The table has three T-slots, is 200mm x 70mm (7⁷/₈" x 2¼") and moves 134mm x 46mm (5¼" x 1¹³/₁₆"). Vertical-head travel is 80mm (3¹/₈"). The mill measures 130mm x 225mm (5¹/₈" x 8⁷/₈"), is 340mm (42½") high and weighs just 7kg (15½lb).

The **Micro Miller FF 230** is a small mill. The 140W (¹/₅hp) mains motor gives six spindle speeds from 280 - 2,200rpm via pulleys and a Poly-V belt. The head swivels through 360° and collets of 6, 8 and 10mm (¼", ⁵/₁₆" and ³/₈") are included. The T-slotted table is 270mm x 80mm (10²/₃" x 3¹/₈") and its travel is 170mm x 65mm (7" x 2½"). The overall height of the mill is 500mm. It weighs 17kg (37½lb).

The **Micro Miller FF 400** uses a 400W (½hp) single-phase reversible motor to give six spindle speeds from 140 - 2,470rpm via a Poly-V belt and pulleys. The headstock swivels through 360°. The quill travel is 30mm (1¹/₈") and the throat distance 100mm (4"). A set of metric collets

Figure 10. *The Sherline 5400 vertical-milling machine. Photo courtesy Sherline.*

is included. The 400mm x 125mm (15¾" x 5") table has three T-slots and its travel is 225mm x 80mm (8⁷/₈" x 3¹/₈"). The mill measures 370mm x 350mm (14½" x 13¾"), is 600mm (23⁵/₈") high and weighs 40kg (88lb).

The **FF 500 Micro Mill** also has a 400W (½hp) single-phase 220/240V motor that gives six spindle speeds from 180 - 2,500rpm via a Poly-V belt drive. The mill head is fitted to a column with dovetail slideways and can be pivoted 360°. Head vertical movement is 220mm (8²/₃"). It comes with four metric collets. The table measures 400mm x 125mm (15¾" x 5") with three T-slots and moves 310mm x 100mm (12¼" x 4"). The mill is 780mm (30¾") high and weighs 47kg (103lb).

Sherline

There are three basic models of Sherline mill, the **5000**, the **5400** and the **2000**.

The **Model 5000** (**5100** metric) features a solid 10" (254mm) aluminium base,

49

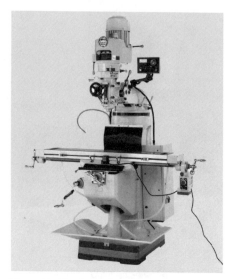

Figure 11. *The South Bend 9 x 42 milling machine. Photo courtesy South Bend Lathe Co.*

precision-machined dovetailed slides with adjustable gibs, permanently lubricated spindle bearings, adjustable pre-load, anti-backlash feed screws on X and Y axes, two 1⁵/₈" (41mm) laser-engraved aluminium hand wheels and one 2½"

(63mm) laser-engraved hand wheel with thrust bearings.

The **Model 5500** (**5510** metric) is in essence the same machine but it is fitted with upgraded hand wheels that feature an adjustable "zero".

The **Sherline Model 5400** (**5410** metric) is an upgraded model with a 12" (305mm) solid-aluminium base, laser-engraved X-axis scales, 2" (51mm) hand wheels on the X and Y axes, a 2½" (63mm) zero-adjustable hand wheel with ball-bearing thrust on the Z axis, a head spacer to increase throat distance and a drill chuck.

Sherline's **Model 2000** (**2010** metric) has a column that offers four additional directions of movement compared to the **Model 5000** and **5400** series mills. The base is extended to 14" (356mm) long to accommodate the additional mechanism of the column. A ¼" (6mm) drill chuck and key are included. Holes can be drilled or surfaces milled from almost any angle with the part mounted flat on the table.

South Bend

The **South Bend 9 x 42 mill** comes as an 8-speed, 16-speed or variable-speed milling machine. All are turret mills with a

Details	5000 (5100)	5400 (5410)	2000 (2010)
Max distance table to spindle	8" (203mm)	8" (203mm)	9" (229mm)
Throat	2¼" (50mm)	2¼" (50mm)	Adjustable
With headstock spacer	Not included	3½" (89mm)	Not required
Travel, "X" Axis	9" (228mm)	9" (228mm)	9" (229mm)
Travel, "Y" Axis	3" (76mm)	5" (127mm)	7" (178mm)
Travel, "Z" Axis	6¼" (159mm)	6¼" (159mm)	5³/₈" (137mm)
Spindle hole	²/₅" (10mm)	²/₅" (10mm)	²/₅" (10mm)
Width overall	14¾" (375mm)	15" (381mm)	15" (381mm)
Depth overall	11¾" (298mm)	14" (356mm)	22¼" (565mm)
Height overall	20¾" (527mm)	20¾" (527mm)	23³/₈" (568mm)
Table size	2¾" x 13" (70 x 330mm)	2¾" x 13" (70 x 330mm)	2¾" x 13" (70 x 330mm)
Shipping weight	33lb (15kg)	36lb (16.3kg)	38lb (17.2kg)

Table 1. *The sizes and weights of the three different Sherline vertical-milling machines.*

knee and a ram. Powered by a 2hp (1.5kW) single-phase motor, they are pre-wired for 220V and, depending on the model, either provide speeds of:
80 - 2,760rpm in eight steps.
90 - 5,600rpm in sixteen steps.
Variable from 60 - 4,500rpm.
The head swivels 90° each side, 45° fore and aft and the over arm can swivel though 360°. Over-arm travel (in/out) is 12" (305mm) and the distance from the spindle nose to the table can vary from 2¾" - 18¾" (70 - 476mm). The table size is 42" x 9" (1,067 x 229mm) and its travel is 27" x 13" (686 x 330mm). Vertical (knee) travel is 16" (406mm) and quill travel is 5" (127mm). The mill measures 81" x 64" x 86¾" (2,057 x 1,625 x 2,203mm) and it weighs 1,924lb (875kg). Standard items include longitudinal table power feed with rapid override, a work light, one-shot lubrication and a toolbox.

The larger **South Bend 9 x 48 EVS** has a 3hp (2.2kW) three-phase variable-speed motor driven by a Yaskawa inverter, which operates from a single-phase 220v supply and provides variable spindle speeds from 50 - 5,000rpm. The head swivels 90° sideways and 45° fore and aft with an over-arm swivel of 360°. Over-arm travel (in/out) is 15" (381mm). Vertical (knee) travel is 16" (406mm) and quill travel is 5" (127mm). The table measures 48" x 9" (1,219 x 229mm) and can travel 35" x 13" (889 x 330mm). The distance from the spindle nose to the table varies from 2¾" - 18¾" (70 - 476mm). The mill measures 99" x 66½" x 87" (2,514 x 1,689 x 2,209mm) and weighs 2,084lb (947kg). The mill has longitudinal table power feed, one-shot lubrication, a coolant system, a chip pan, a work light and a toolbox.

Taig/Peatol Micro Mill
There are two different versions of the American **Taig Micro Mill**; one with a

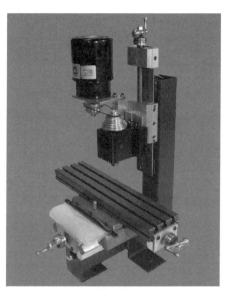

Figure 12. *The Taig/Peatol Micro Mill. Photo courtesy Peatol Machine Tools.*

larger table than the other, as well as a CNC version described in Chapter 4. Taig sell in the UK under the Peatol brand. The smaller **2018ER** has a 3½" x 15¾" (89mm x 400mm) table with 9½" (241mm) of travel. The **2019ER** has a 3½" x 18½" (89mm x 470mm) table and 12" (305mm) of X-axis travel. Travel for both machines in the Y axis is 5½" (140mm) and in the Z axis is 6" (152mm).

For both machines, the ⅛hp (150W) motor provides six spindle speeds from 525 - 5,200rpm and the spindle head can swivel and rotate 90° each way. The maximum distance from the spindle to the table is 9" (229mm) and it will take collets up to ³/₈" (9.5mm). The mill measures 17" x 16¾" (432mm x 425mm), is 17" high (432mm) and weighs 65lb (29.5kg). The makers state that the overall working accuracy of both the mills should exceed 0.0005" (0.013mm).

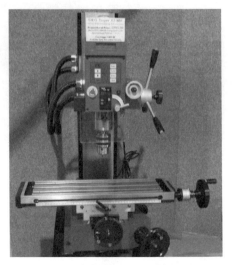

Figure 14. *The Sieg Super X3 mill, typical of many Far-East bench-mounted machines, on display at a model engineering exhibition.*

Figure 13. *The Wabeco F1210E bench-top mill.*

Wabeco

Made in Germany, Wabeco build a range of six vertical-milling machines.

The **Wabeco F1200 E, F1210 E** and **F1410 LF** mills are fitted with single-phase 1.4kW (1^7/$_8$hp) motors suitable either for 230 or 110volts with electronic infinitely variable spindle speeds from 140 - 3,000rpm. The milling head is able to swivel 90° either way. Vertical movement is 280mm (14½") up and down and the quill stroke is 55mm (2"). The distance from the spindle nose to the table varies from 350mm to 370mm (13¾" to 14^2/$_3$") and this depends on the model. An optional base cabinet houses a coolant unit for the **F1200 E** and **F1210 E**.

The **F1200 E** milling table measures 450 x 180mm (17¾" x 7") with automatic feed in all three axes; those of the **F1210 E** and

Machine	Dimensions (W x D x H)		Weight	
F1200 E	650 x 610 x 670mm	(25½" x 24" x 26^1/$_3$")	85kg	(187lb)
F1210 E	900 x 610 x 670mm	(35½" x 24" x 26^1/$_3$")	101kg	(222lb)
F1410 LF	915 x 700 x 780mm	(36" x 27½" x 30¾")	118kg	(260lb)
F1200 E*	650 x 610 x 870mm	(25½" x 24" x 34¼")	96kg	(211lb)
F1210 E*	900 x 610 x 870mm	(35½" x 24" x 34¼")	112kg	(246lb)
F1410 LF*	915 x 860 x 870mm	(36" x 33^7/$_8$" x 34¼")	127kg	(279lb)

Table 2. *The sizes and weights of the various Wabeco vertical-milling machines. * = high speed.*

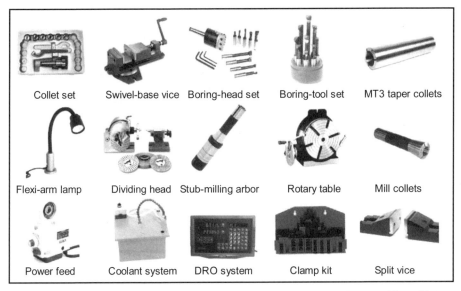

| Collet set | Swivel-base vice | Boring-head set | Boring-tool set | MT3 taper collets |

| Flexi-arm lamp | Dividing head | Stub-milling arbor | Rotary table | Mill collets |

| Power feed | Coolant system | DRO system | Clamp kit | Split vice |

Figure 15. *The optional accessories typically available for many Far-East milling machines. Photos courtesy Chester.*

F1410 LF are 700 x 180 mm (27½" x 7"). Each table has three T-slots. The **F1200 E** can be moved 260mm (15") sideways (the **F1210 E** and **F1410 LF** 500mm [19²/₃"]). The **F1200 E** and **F1210 E** can move fore and aft 150mm (6") or optionally 180mm (7"); the **F1410 LF** a total of 200mm (7⁷/₈"). Alternative high-speed versions of each of these three models has a 2kW (2³/₃hp) motor and a spindle speed range from 100 to 7,500rpm. Their dimensions and weights are shown in Table 2. Both metric and imperial versions are available.

Far East vertical-milling machines
The same reservations mentioned in Chapter 1 about lathes made in the Far East also apply to milling machines. In the UK, Amadeal, Arc Euro Trade, Axminster, Chester, Clarke International, Excel Machine Tools, Machine Mart, RDG

Tools and Warco all sell a wide range of imported machines. They are amazingly low-cost when compared to any similar machines designed and manufactured in the industrialised nations.

Similar comments to those on page 17 also apply to the Opti and Quantum mills that are imported in the UK by Excel Machine Tools.

Many Far-East built machines have very similar specifications and most are available in metric or imperial format. They are all dual-purpose machines with a fine screw control for setting the milling cutter depth and a coarse hand-quill feed for drilling.

Similar imported machines are offered by many American suppliers such as All Machine Tools, Bolton Hardware, Grizzly Industrial Inc, Harbor Freight Tools, LatheMaster Metalworking Tools, Southern Tool and Sears.

53

Mill	Table size	Table movement	Vertical travel	Quill travel
Axminster Sieg Super X1	240 x 145mm (9½" x 5¾")	180 x 90mm (7 x 3½")	-	30mm (1¹/₈")
Chester Cobra	240 x 145mm (9¾" x 5¾")	180 x 145mm (7" x 5¾")	225mm (9")	30mm (1¹/₈")
Warco Micro Mill	240 x 145mm (9¾" x 5¾")	180 x 145mm (7" x 5¾")	225mm (9")	-
Clarke CMD10	240 x 145mm (9¾" x 5¾")	180 x 145mm (7" x 5¾")	225mm (9")	-
Axminster SIEG X2 and Chester Conquest	390 x 92mm (15¹/₃" x 3²/₃")	100 x 180mm (4" x 7")	180mm (7")	-
Clarke CMD300	400 x 92mm (15¾" x 3²/₃")	187 x 100mm (7¹/₃" x 4")	-	-
Warco WM-14	400 x 120mm (16" x 4½")	160 x 220mm (6⁵/₁₆" x 8¹¹/₁₆")	210mm (8⁵/₁₆")	50mm (2")
Quantum BF 16 Vario	400 x 120mm (15¾" x 4¾")	220 x 160mm (8²/₃" x 6¼")	210mm (8¼")	50mm (2")
RDG 16VSM	400 x 120mm (15¾" x 4¾")	230 x 145mm (9" x 5¾")	210mm (8¼")	50mm (2")
Axminster SIEG U2	450 x 120mm (17¾" x 4¾")	120mm (4¾")	-	-

Table 3a. *The range of Far-East-manufactured mills sold in the UK, in order of table size.*

The mills listed in Table 3a suit the smaller workshop and the first four mills all have the same size tables.

The **Axminster Sieg Super X1** is the lightest machine with a table that can be moved in two directions. It has a pair of electronically variable speed ranges and is supplied with a 10mm (³/₈") chuck and a set of service tools. However, it is only available in metric form.

Slightly smaller than the **Sieg Super X1** but heavier and with greater table movement, the **Chester Cobra** and the **Warco Micro Mill** seem to be re-badged versions of the same machine with only some very minor differences. Both have gear-driven electronically controlled variable-speed spindles. The **Cobra** mill comes with a 10mm (³/₈") drill chuck, a milling drawbar and a set of basic service tools.

A small dovetail-column mill, the **Clarke CMD10** appears very similar but lacks any head-tilt facility.

The next seven mills have significantly larger tables. Two, the **Axminster SIEG X2 Mini Mill** and the **Chester Conquest** have the same specifications though the Axminster mill is only available in metric form. They feature a two-speed gearbox and are fitted with a 13mm (½") drill chuck. They are capable of 13mm (½") drilling, 16mm (²/₃") end milling and 30mm (1¹/₈") face milling.

The **Clarke CMD300** is bench mounted and has a slightly larger table than the previous two mills, as do the **Warco WM-14** and the **Quantum BF 16 Vario**, that have obviously been designed to quite similar specifications. The **CMD300** head only tilts 45° each way while the heads of

Spindle to table	Head tilt	Motor	Speeds in rpm	Length x width x height	Weight
255mm (10")	45°	150W ($^1/_5$hp)	0 - 1,000 0 - 2,000	430 x 355 x 715mm (17" x 14" x 28")	32kg (70lb)
220mm (8¾")	45°	150W ($^1/_5$hp)	100 - 2,000	425 x 350 x 690mm (17" x 13¾" x 27")	40kg (88lb)
266mm (10½")	45°	150W ($^1/_5$hp)	100 - 2,000	425 x 350 x 690mm (17" x 13¾" x 27")	40kg (88lb)
140mm (5½")	None	150W ($^1/_5$hp)	100 - 2,000	425 x 350 x 690mm (17" x 13¾" x 27")	40kg (88lb)
170mm (6$^2/_3$")	45°	350W (½hp)	0 - 1,100 0 - 2,500	520 x 500 x 760mm (20½" x 19$^2/_3$" x 30")	50kg (110lb)
-	45°	470W ($^2/_3$hp)	0 - 2,500	514 x 506 x 756mm (20¼" x 20" x 29¾")	51kg (112lb)
280mm (11")	90°	500W ($^2/_3$hp)	500 - 2,250	500 x 460 x 800mm (19$^2/_3$" x 18" x 31½")	60kg (132lb)
285mm (11¼")	90°	500W ($^2/_3$hp)	100 - 3,000	510 x 450 x 760mm (20" x 17¾" x 30")	60kg (132lb)
-	90°	500W ($^2/_3$hp)	50 - 2,250	480 x 450 x 820mm (19" x 17¾" x 32¼")	60kg (132lb)
-	90°	1kW (1$^1/_3$hp)	200 -2,500	570 x 520 x 1,600mm (22½" x 20½" x 63")	185kg (407lb)

the other two can be rotated the full 90° in either direction.

The **RDG 16VSM** geared-head mill has a Jacobs chuck with arbor and drawbar, a digital depth readout and a toolbox with a set of tools.

The **Axminster SIEG U2** Mill is an unusual milling machine as it also offers a surface-grinding capability with a fixed grinding speed of 2,500rpm. This can be a very desirable facility. The machine has a much more powerful motor than the other machines in its class and is supplied with a grinding attachment and grinding wheel as well as a slitting attachment, a 16mm ($^5/_8$") drill chuck and a floor stand with a storage compartment. This machine is significantly heavier than other machines of a similar physical size; an important factor in optimising grinding performance.

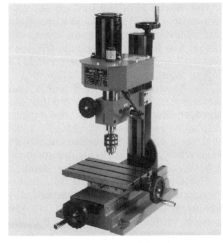

Figure 16. *The Warco Micro Mill. Photo courtesy Warco.*

Mill	Table size	Table movement	Vertical travel	Quill travel
Warco Mini Mill/Drill	460 x 112mm ($18^1/_8$" x $4^2/_8$")	300 x 130mm ($11^3/_4$" x 5")	220mm ($8^2/_3$")	-
Axminster SIEG X2 Super	500 x 130mm ($19^2/_3$" x $5^1/_8$")	250 x 160mm ($9^7/_8$" x $6^1/_4$")	60mm ($2^1/_3$")	-
Chester Champion 16V	500 x 140mm ($19^2/_3$" x $5^1/_2$")	280 x 160mm (11" x $6^1/_4$")	200mm ($7^7/_8$")	52mm (2")
Opti BF 20 Vario	500 x 180 mm ($19^2/_3$" x 7")	280 x 175mm (11" x $6^7/_8$")	280mm (11")	50mm (2")
Axminster SIEG X3	550 x 160mm ($21^2/_3$" x $6^1/_4$")	400 x 145mm ($15^3/_4$" x $5^3/_4$")	-	80mm ($3^1/_8$")
Axminster SIEG Super X3	550 x 160mm ($21^2/_3$" x $6^1/_4$")	400 x 145mm ($15^3/_4$" x $5^3/_4$")	-	-
Chester Lux Pedestal	580 x 190mm ($22^7/_8$" x $7^1/_2$")	420 x 170mm ($16^1/_2$" x $6^2/_3$")	-	130mm ($5^1/_8$")
Warco Economy Mill/Drill	585 x 190mm (23" x $7^1/_2$")	355 x 150mm (14" x 4")	-	100mm (4")
Clarke CMD1225C	585 x 190mm (23" x $7^1/_2$")	-	-	-
Chester Eagle E25	585 x 190mm (23" x $7^1/_2$")	370 x 140mm ($14^1/_2$" x $5^1/_2$")	-	100mm (4")

Table 3b. *The range of Far-East-manufactured mills sold in the UK, in order of table size.*

The **Warco Mini Mill/Drill** comes with two variable speed ranges and a back gear that will maximise torque in the low-speed range. It is fitted with a spindle fine feed and a 12.5mm (½") drill chuck on an MT3 Morse taper.

The **Axminster Sieg X2 Super** is a far larger and more capable machine than the standard **Sieg X2**. It is fitted with a digital display of both spindle speed and travel and has a 13mm (½") keyless chuck with MT2 arbor. It is suitable for mounting on a bench or on the optional floor stand.

The **Chester Champion 16V** has a marginally larger table but is significantly lighter. It has a gear head to provide greater torque in the lower of the two electronically controlled speed ranges.

The **Opti BF 20 Vario** has a two-stage, continuously variable speed range, a digital sleeve-travel indicator reading to 0.01mm (0.0004") and a machine lamp. It has a drilling capacity of 16mm ($^2/_3$") and can use end mills with a diameter of up to 20mm (¾").

Both the **Axminster SIEG X3** and **Axminster SIEG Super X3** have very similar specifications and on both versions the vertical-feed control wheel is located easily to hand on the base. The **Sieg X3** has two electronically variable spindle speed ranges. Only the **Super** can tilt its milling head. It has a significantly more powerful motor, both digital spindle-speed and down-feed read-outs as well as a thread-tapping facility.

The next four machines all have round columns and no electronic-speed control. **Chester Lux Pedestal** is the only one of the four that provides head tilt. It has a

Spindle to table	Head tilt	Motor	Speeds in rpm	Length x width x height	Weight
290mm (11½")	45°	550W (¾hp)	50 - 1,100 / 50 - 2,500	670 x 510 x 780mm (26$^{1}/_{3}$" x 20½" x 30¾")	62kg (136lb)
320mm (12$^{2}/_{3}$")	45°	700W (1hp)	50 - 2,500	610 x 610 x 780mm (24" x 24" x 37¾")	127kg (279lb)
285mm (11¼")	90°	600W (¾hp)	50 - 1,250 / 100 - 2,500	550 x 530 x 760mm (21$^{2}/_{3}$" x 20$^{7}/_{8}$" x 30")	90kg (198lb)
185mm (7¼")	90°	850W (1$^{1}/_{8}$hp)	90 - 3,000	670 x 550 x 860mm (26$^{1}/_{3}$" x 21$^{2}/_{3}$" x 33$^{7}/_{8}$")	103kg (227lb)
380mm (15")	None	600W (¾hp)	100 - 1,000 / 100 - 2,000	685 x 360 x 830mm (27" x 14$^{1}/_{8}$" x 32$^{2}/_{3}$")	165kg (363lb)
380mm (15")	90°	1kW (1$^{1}/_{3}$hp)	100 - 1,750	685 x 560 x 830mm (27" x 22" x 32$^{2}/_{3}$")	165kg (363lb)
680mm (26¾")	90°	1.1kW (1½hp)	95 - 1,600	960 x 700 x 1,460mm (37¾" x 27½" x 57½")	280kg (616lb)
380mm (15")	None	750W (1hp)	120 - 2,580	940 x 965 x 940mm (37" x 38" x 37")	185kg (407lb)
-	None	750W (1hp)	100 - 2,150	940 x 900 x 940mm (37" x 35½" x 37")	167kg (367lb)
200mm (8")	None	550W (¾hp)	100 - 2,150	900 x 940 x 1,040mm 35½" x 37" x 41")	200kg (440lb)

gear head that provides a choice of six different speeds using lever control. It comes with a drill chuck, an arbor, a milling drawbar and a V100 machine vice with a swivel base.

The **Warco Economy Milling/Drilling Machine** has been available for 30 years and features eleven belt-driven spindle speeds. The head is mounted on a 92mm (3$^{5}/_{8}$") diameter column. An optional stand with built-in coolant tray is available.

The **Clarke CMD1225C** is a metric mill with a round column, has a head that does not tilt and has twelve spindle speeds. There is an optional stand with a cupboard.

The **Chester Eagle E25 Mill/Drill** uses twelve spindle speeds for a head that can swivel 360° around the column but does not tilt. It has a 13mm (½") drill chuck, an arbor, a drawbar and a small tilting vice.

Figure 17. *The Clarke CMD 1225C vertical-milling machine. Photo courtesy Chronos.*

Mill	Table size	Table movement	Vertical travel	Quill travel
Axminster ZX25M2	585 x 190mm (23" x 7½")	370 x 160mm (14½" x 5½")	- -	100mm (4")
Chester Century and Amadeal XJ20	600 x 180mm (23²/₃" x 7")	350 x 200mm (13¾" x 7⅞")	400mm (15¾")	60mm (2¹/₃")
Warco ZX-15	630 x 150mm (25" x 6")	350 x 145mm (13¾" x 5¾")	125mm (5")	85mm (3³/₈")
Warco WM-16 and Amadeal AMA 25V	700 x 180mm (27½" x 7")	485 x 175mm (19" x 6⁷/₈")	380mm (15")	50mm (2")
Chester Champion 20V	700 x 180mm (27½" x 7")	490 x 290mm (19¼" x 11¹/₃")	290mm (11½")	42mm (1²/₃")
Opti BF 20 L Vario	700 x 180mm (27½" x 7")	280 x 175mm (11" x 6⁷/₈")	280mm (11")	50mm (2)"
RDG 20VSM	700 x 180mm (27½" x 7")	480 x 145mm (18⁷/₈" x 5¾")	360mm (14¹/₈")	50mm (2)"
Warco WM-18V	700 x 210mm (27½" x 9¼")	425 x 220mm (16¾" x 8⁵/₈")	370mm (14½")	50mm (2")

Table 3c. *The range of Far-East-manufactured mills sold in the UK, in order of table size.*

Of the mills listed on this page, only the first and the fourth have a round column.

The **Axminster ZX25M2** has twelve belt-driven speeds and a 13mm (½") chuck with arbor. It is only available in metric form and has the useful option of a longitudinal power-feed unit.

The **Chester Century** is a variable-speed machine with a cast-iron pillar and

Figure 18. *The Warco WM-16 head will tilt 90° sideways. Photo courtesy Warco.*

a dovetailed slide. A gear head gives two spindle-speed ranges to a 20mm (⁷/₈") drill chuck. There is a digital quill display and under-head work light. The **Amadeal XJ20** appears to be identical.

The **Warco ZX-15 Milling Machine** (illustrated in Figure 3 on page 45 has a 70mm (2¼") diameter column allowing the head to swivel right around the column through 360° and has a 13mm (½") drill chuck. A separate swarf tray and stand with built-in tray are offered. The mill can be purchased in imperial or metric form.

The **WM-16 Variable Speed Mill** has a reversible motor giving infinitely variable speeds. At low speed, a back gear will increase torque. A digital rev counter and a depth gauge are fitted. It comes with a drill chuck and an arbor with a draw bar. Options include a separate swarf tray or a stand with a built-in tray. The purchaser can choose an imperial or metric mill.

The **Amadeal AMA 25V** is similar but has 60mm (2¹/₃") less sideways table movement. A variant has an X-axis power feed.

Spindle to table	Head tilt	Motor	Speeds in rpm	Length x width x height	Weight
380mm (15")	None	750W (1hp)	90 - 2,150	1,320 x 900 x 960mm (52" x 35½" x 37¾")	220kg (484lb)
175mm (6⁷/₈")	90°	1.1kW (1½hp)	50 - 3,000	720 x 565 x 1,020mm (28¹/₃" x 22¼" x 40¹/₈")	135kg (297lb)
400mm (16")	90°	550W (¾hp)	400 -1,640	775 x 559 x 1,067mm (30½" x 22" x 42")	135kg (297lb)
-	90°	600W (⁴/₅hp)	50 - 2,250	355 x 432 x 813mm (14" x 17" x 32")	103kg (227lb)
390mm (15¹/₃")	90°	750W (1hp)	50 - 2,200	960 x 570 x 970mm (37¾" x 22½" x 38¹/₈")	113kg (249lb)
185mm (7¼")	90°	850W (1¹/₈hp)	90 - 3,000	870 x 550 x 860mm (34¼" x 21²/₃" x 33⁷/₈")	115kg (253lb)
-	90°	600W (⁴/₅hp)	50 - 2,200	760 x 690 x 860mm (30" x 27¹/₈" x 33⁷/₈")	115kg (253lb)
-	90°	1.1kW (1½hp)	50 - 3,000	991 x 711 x 1,042mm (39" x 28" x 41")	220kg (484lb)

The **Chester Champion 20V** is available in metric or imperial form with digital speed and quill readouts. With a 750W (1hp) motor, it appears to be a very comparable machine to the **Warco WM-16**. It has a variable speed control and comes with a drill chuck, an arbor and draw bar. An optional floor stand is also available.

The **Opti BF 20 L Vario** is the same as the **Opti BF 20 Vario** but with a 200mm (7⁷/₈") longer table that has 70mm (2¾") more lateral movement. It also has two continuously variable speed ranges, a digital sleeve-travel indicator as well as a machine lamp. It has a drilling capacity of 16mm (²/₃") and can use end mills up to 20mm (¾") diameter.

The **RDG 20VSM Mill** provides variable speeds and has a Jacobs chuck with an arbor and drawbar, digital depth and speed readouts and a toolbox with tools.

The **Warco WM-18VS Mill** comes with an infinitely variable range of spindle speeds to a head, fitted with a 13mm (½") drill chuck. A back gear gives maximum torque at low speed. A digital speed readout and depth gauge are standard fittings. Options include a custom swarf tray or a stand with a built-in tray. There is a choice of an imperial or a metric mill.

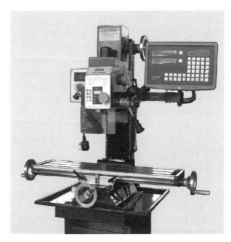

Figure 19. *The RDG 20VSM with optional Sino digital readout. Photo courtesy RDG Tools.*

Mill	Table size	Table movement	Vertical travel	Quill travel
RDG 30VSM	700 x 210mm (27½" x 8¼")	430 x 200mm (17" x 7⅞")	380mm (15")	70mm (2¾")
Axminster RF31	721 x 210mm (28⅓" x 8¼")	520 x 178mm (20½" x 7")	-	125mm (5")
Axminster ZX30M	730 x 210mm (28¾" x 8¼")	500 x 175mm (19⅔" x 6⅞")	-	130mm (5⅛")
Chester Eagle 30	730 x 210mm (28¾" x 8¼")	540 x 180mm (21¼" x 7")	-	130mm (5⅛")
Opti F 30	730 x 210mm (28¾" x 8¼")	430 x 185mm (17" x 7¼")	465mm (18⅓")	130mm (5⅛")
Axminster RF40	730 x 210mm (28¾" x 8¼")	500 x 230mm (19⅔" x 9")	-	130mm (5⅛")
Chester Lux Round Column	730 x 210mm (28¾" x 8¼")	540 x 180mm (21¼" x 7")	-	130mm (5⅛")
Warco Major GH	730 x 210mm (28¾" x 8¼")	500 x 270mm (19⅔" x 10¼")	-	130mm (5⅛")
Warco Major	736 x 210mm (29" x 8¼")	508 x 203mm (20" x 8")	-	130mm (5⅛")

Table 3d. *The range of Far-East-manufactured mills sold in the UK, in order of table size.*

The nine machines shown in the Table 3d above are mostly very similar in size as well as in weight.

There are two variants of the **RDG 30VSM Mill**. The second one has a wider table of 840mm (33") with an extra travel of 570mm (22½") resulting in an increased width of 850mm (33½") and a heavier weight of 215kg (473lb). Both can provide variable speeds and have a Jacobs chuck with an arbor and drawbar, digital depth and speed readouts and a toolbox with tools.

The **Axminster RF31 Vertical Mill/Drill** has a single-phase motor giving twelve belt-driven spindle speeds. It can be mounted on a suitable bench or on the optional floor stand. An optional power feed will provide longitudinal travel. The mill is illustrated in Figure 20.

The **Axminster ZX30M Mill/Drill** also has a motor that gives twelve speeds via a belt drive to the spindle fitted with a 13mm (½") drill chuck. The head is mounted on a circular column allowing it to swivel 360°. The mill can be mounted on a sturdy bench or on an optional floor stand. It is available in metric form only. An optional power-feed unit will provide longitudinal travel.

The **Chester Eagle 30** has a similar specification. Its head can be rotated through 360° and it comes with a 13mm (½") drill chuck, a tilting vice, a face-mill cutter and a draw bar. The wide range of accessories includes a coolant system and a floor stand. It is available either in metric or imperial form.

The **Opti F30** is offered either in single- or three-phase versions while the **Opti F 30 Vario** is only a three-phase machine. The Vario has electronic speed control and is 5kg (11lb) heavier. Both mills come with a 13mm (½") chuck and a set of tools.

The **Axminster RF40 Universal Mill** is quite similar to the **RF31** but is fitted with a six-speed gearbox. It is also similar to

Spindle to table	Head tilt	Motor	Speeds in rpm	Length x width x height	Weight
-	90°	1.1kW (1½hp)	50 - 2,250	755 x 760 x 1,160mm (29¾" x 30" x 45²/₃")	200kg (440lb)
457mm (18")	None	1.2kW (1½hp)	125 - 2,500	1,118 x 940 x 1,092mm (44" x 37" x 43")	300kg (660lb)
202mm (8")	None	1.5kW (2hp)	100 - 2,080	1,095 x 1,010 x 1,125mm (43" x 39¾" x 44¼")	270kg (594lb)
460mm (18")	None	1.1kW (1½hp)	100 - 2,150	1,120 x 1,010 x 1,080mm (44" x 39¾" x 42½")	287kg (631lb)
-	None	1.5kW (2hp)	125 - 2,500 (Vario 35 - 3,300)	1,120 x 1,100 x 1,100 (44" x 43¹/₃" x 43¹/₃")	270kg (594lb)
470mm (18½")	90°	1.2kW (1½hp)	50 - 1,250	800 x 1,100 x 1,100mm (31½" x 43¹/₃" x 42¹/₃")	300kg (660lb)
460mm (18")	45°	750W (1hp)	95 - 1,600	960 x 700 x 1,460mm (37¾" x 27½" x 57½")	280kg (616lb)
470mm (18½")	90°	1.1kW (1½hp)	75 - 1,600	1,067 x 762 x 1,422mm (42" x 30" x 56")	320kg (704lb)
480mm (19")	None	1.1kW (1½hp)	100 - 2,080	1,067 x 1,016 x 1,078 (42" x 40" x 42½")	300kg (660lb)

the **Chester Lux Round Column mill/ drill**, which has a slightly larger table but a less powerful motor, and the **Warco Major GH** and the **Warco Major Milling Drilling Machine**. All three of these mills have a higher top speed, particularly the last one of these machines.

The **Chester Lux Round Column** comes with a 13mm (½") chuck, a swivel-base machine vice and a face-mill cutter.

The **Warco Major GH** has a gear driven spindle with six lever-controlled speeds. It employs a 114mm (4½") round column that allows the head to swivel through 360°. It is fitted with a 13mm (½") chuck. An optional wide tray and a floor stand are both available.

The **Warco Major Milling Drilling** is a round-column machine that uses a belt drive to provide twelve spindle speeds. It can be supplied in metric or imperial form. An optional stand, a tray and a power table feed are available.

Figure 20. *The Axminster RF31 vertical mill. Photo courtesy Axminster Power Tool Centre.*

Mill	Table size	Table movement	Vertical travel	Quill travel
Opti BF 30 Vario	750 x 210mm (29½" x 8¼")	450 x 200mm (17¾" x 7⁷/₈")	470mm (18½")	90mm (3½")
Warco GH Universal	800 x 240mm (31½" x 9½")	585 x 220mm (23" x 8²/₃")	-	130mm (5¹/₈")
Warco Super Major	800 x 240mm (31½" x 9½")	585 x 220mm (23" x 8²/₃")	-	130mm (5¹/₈")
Chester Lux	820 x 240mm (32¼" x 9½")	520 x 190mm (20½" x 7½")	-	130mm (5¹/₈")
Chester Super Lux	820 x 240mm (32¼" x 9½")	530 x 200mm (20⁷/₈" x 7⁷/₈")	-	120mm (4¾")
Opri F 45	820 x 240mm (32¼" x 9½")	520 x 210mm (20½" x 8¼")	510mm (20")	130mm (5¹/₈")
Opti BF 46 Vario	850 x 240mm (33½" x 9½")	500 x 250mm (19²/₃" x 9⁷/₈")	541mm (21¼")	115mm (4½")

Table 3e. *The range of Far-East-manufactured mills sold in the UK, in order of table size.*

The **Opti BF 30 Vario** is a totally new machine. Unlike the **Opti F 30 Vario** it does not have a round column. Its spindle is gearbox driven giving three ranges of variable speeds with digital-speed and spindle-position readouts. A floor stand is available as an optional extra.

The **Warco GH Universal Milling Drilling Machine** has a square column to support its head and has a geared-crank-handle down feed. A gearbox allows lever-control change of spindle speeds. Warco offer an optional coolant system, a floor stand and many other accessories.

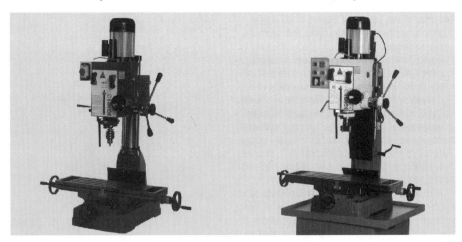

Figure 21. *The Chester Lux Round Column (left) is a powerful bench-mounted milling machine. The Chester Super Lux (right) has an entirely different type of column. Photos courtesy Chester.*

Spindle to table	Head tilt	Motor	Speeds in rpm	Length x width x height	Weight
-	90°	1.5kW (2hp)	35 - 3,300	1,073 x 792 x 1,245mm (42¼" x 31⅛" x 49")	265kg (583lb)
450mm (17¾")	45°	1.1kW (1½hp)	75 - 1,600	1,117 x 787 x 1,410mm (44" x 31" x 55½")	320kg (704lb)
450mm (17¾")	45°	1.1kW (1½hp)	75 - 1,600	1,245 x 915 x 2,134mm (49" x 36" x 84")	450kg (990lb)
470mm (18½")	90°	750W (1hp)	95 - 1,600	880 x 780 x 1,150mm (34⅔" x 30⅔" x 45¼")	300kg (660lb)
470mm (18½")	90°	750W (1hp)	95 - 1,600	1,120 x 1,040 x 2,100mm (44" x 41" x 82⅔")	278kg (612lb)
-	90°	1.1kW (1½hp)	50 - 2,560	785 x 1,185 x 1,450mm (31" x 46⅔" x 57")	330kg (726lb)
-	90°	2.2kW (3hp)	65 - 3,300	1,230 x 948 x 1,519mm (48½" x 37⅓" x 59⅞")	480kg (1,056lb)

The **Warco Super Major Milling Drilling Machine**, although 130kg (286lb) heavier, is of a similar design to the **GH Universal**. The vertical movement of its head and table longitudinal travel are power operated. The mill comes complete with 13mm (½") chuck and a stand. A wide range of accessories is available.

The **Chester Lux** has six gear-driven spindle speeds. It comes with a 13mm (½") chuck, a face-mill cutter, and a machine vice with swivel base. Options include a stand, a coolant system, a flexiarm lamp and a longitudinal table power-feed system.

The **Chester Super Lux** is quite similar to the **Chester Lux** but has a powered elevation head mounted on a dovetailed column and includes a floor stand. It has a drilling capacity of 45mm (1¼"), end mill capability of 28mm (1⅛") and a face mill capacity of 80mm (3⅛").

The **Opti F 45** has a two-step motor that gives just 500W (⅔hp) at the low setting. It can drill and end mill up to 32mm (1¼") and face mill up to 100mm (4"). It includes a cutter head with holding taper, draw-in rod and operating tools.

The **Opti BF 46 Vario** has a three-stage gearbox that provides three ranges of overlapping variable spindle speeds. It can drill 45mm (1¾"), end mill up to 32mm (1¼") and face mill up to 80mm (3⅛"). It has both a digital speed and a depth readout. An optional coolant system and a floor stand, complete with chip drip plates, are available.

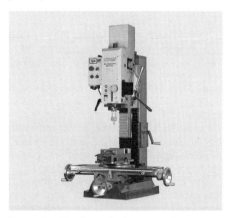

Figure 22. *The Opti BF 46 Vario vertical-milling machine. Photo courtesy Optimum Maschinen.*

Mill	Table size	Table movement	Vertical travel	Quill travel
Warco VMC Turret	660 x 152mm (26" x 6")	370 x 152mm (14½" x 6")	- -	89mm (3½")
Chester 836 Turret	660 x 200mm (26" x 7⁷/₈")	609 x 240mm (24" x 9½")	345mm (13½")	127mm (5")
Chester 626 Turret	745 x 156mm (29¹/₃" x 6¹/₈")	380 x 135mm (15" x 5¹/₃")	330mm (13")	80mm (3¹/₈")
Chester 626S Turret	860 x 250mm (33⁷/₈" x 9⁷/₈")	440 x 290mm (17¹/₃" x 11½")	460mm (18")	100mm (4")
Warco WM-20	910 x 226mm (36" x 9")	580 x 240mm (22¾" x 9½")	350mm (14")	127mm (5")
Opti MF 2 Vario	1,244 x 230mm (49" x 9")	770 x 305mm (30¹/₃" x 12")	406mm (16")	120mm (4¾")
Warco VSE	1,245 x 230mm (49" x 9")	730 x 306mm (28¾" x 12")	406mm (16")	127mm (5")
Warco 4VS	1,245 x 230mm (49" x 9")	730 x 306mm (28¾" x 12")	406mm (16")	127mm (5")

Table 3f. *The range of Far-East-manufactured turret mills sold in the UK, in order of table size.*

The **Warco VMC Turret Mill** is the smallest of the turret mills and has a nine speed v-belt-driven spindle that can rotate through 360° and tilt. It has a traditional elevating knee and a one-shot lubrication system. It includes a halogen low-volt light and an optional digital readout. Metric and imperial versions of the mill are offered.

The **Chester 836 Turret Mill** comes with a single-phase motor that provides variable spindle speeds from 200rpm or a three-phase one giving a minimum speed of 100rpm. The ram travel is 260mm (10¼"). The distance from the spindle to the column varies from 425 - 100mm (16¾" - 4"). The mill comes with one-shot lubrication and a coolant system, a 3-axis digital readout, longitudinal power feed, a vice and a halogen work light.

The **Chester 626 Turret Mill** has a head that swivels ±90° as well as tilting. It can be supplied with a single- or a three-phase motor that provides nine spindle speeds. The maximum distance from the spindle

to the column is 155mm (6¹/₈"). The mill mounts on a small stand with a cupboard and has a central lubrication system and a machine work light. Table power feed and a coolant system are options. Metric and imperial machines are available.

The **Chester 626S Turret Mill** is a rather different machine despite its very similar designation. It has a bigger table, a larger throat depth with a movable ram and a built-in cast base. It also has a larger capacity, is more powerful and is a heavier floor-standing machine.

The **Warco WM-20 Milling Machine** is fitted with one-shot lubrication, a coolant system and halogen low-voltage lighting. The single-phase motor has an infinitely variable range of spindle speeds. The over arm can rotate 360° as well as tilt and its stroke is 250mm (10"). Options include a two-axis power table feed and powered knee-elevation movement as well as a two-axis digital readout.

The **Opti MF 2 Vario** is made from cast iron with hardened slideways. The milling

Spindle to table	Head tilt	Motor	Speeds in rpm	Length x width x height	Weight
165mm (6½")	45°	1.1kW (1½hp)	160 - 2,540	1,092 x 1,016 x 1,702mm (43" x 40" x 67")	415kg (913lb)
380mm (15")	90°	1.1kW (1½hp)	100/200 - 1,300	1,400 x 1,300 x 1,955mm (55" x 51" x 77")	670kg (1,474lb)
330mm (13")	45°	1.1kW (1½hp)	190 - 2,100	1,085 x 990 x 1,710mm (42¾" x 39" x 67⅓")	410kg (902lb)
400mm (15¾")	45°	1.5kW (2hp)	300 - 5,000	1,499 x 1,500 x 1,800mm (59" x 59" x 70⅞")	500kg (1,100lb)
-	90°	1.5kW (2hp)	25 - 1,480	1,321 x 1,177 x 1,930mm (52" x 46½" x 76")	750kg (1,650lb)
-	45°	1.5kW (2hp)	10 - 5,100	1,480 x 1,680 x 2,150mm (52¼" x 66⅛" x 84⅓")	900kg (1,980lb)
469mm (18½")	180°	1.5kW (2hp)	25 - 2,280	1,468 x 1,620 x 2,010mm (57¾" x 63¾" x 79⅛")	960kg (2,112lb)
469mm (18½")	45°	1.5kW (2hp)	50 - 3,500	1,468 x 1,620 x 2,010mm (57¾" x 63¾" x 79⅛")	960kg (2,112lb)

head can rotate through 360° as well as tilting. The spindle can turn either way and speed is continuously variable. An X-axis power feed, a coolant pump and a chip pan are included. There is an optional three-axis digital readout.

The **Warco VSE Turret Mill** is based on the **Warco 4VS** described below. Both machines are heavy and use a single-phase inverter drive to the motor that gives infinitely variable spindle speeds. Each of the mills comes with as coolant system as well as built-in central lubrication. As an optional extra for either machine, a 2-axis digital readout is also available but only metric versions of the mills are offered; there are no imperial variants.

The **Warco 4VS** is a considerably more capable machine and, compared to the **VSE**, additionally comes complete with a power feed to both the table X- and Y-axes as well as to the knee. A low-volt halogen light and electrical overload protection are included. All of these items may be added as options on the **VSE**.

Figure 23. *The Chester 626S is a large and capable turret mill. Photo courtesy Chester.*

Mill	Table size	Table movement	Vertical travel	Quill travel
Harbor Freight 47158	9³/₈" x 5³/₈" (238 x 136mm)	9" x 4" (229 x 102mm)	8½" (216mm)	1¹³/₁₆" (46mm)
Grizzly G8689	15¾" x 3⁵/₈" (400 x 92mm)	7¹¹/₁₆" x 4" (195 x 102mm)	9⁷/₈" (251mm)	5" (127mm)
Harbor Freight 44991	15⁷/₈" x 3¹¹/₁₆" (403 x 94mm)	- -	8½" (216mm)	- -
Bolton XJ9510	16" x 4" (406 x 102mm)	18½" x 4" (470 x 102mm)	7" (178mm)	- -
Grizzly G0463	21⁵/₈" x 6¼" (549 x 159mm)	15⁷/₈" x 5¼" (403 x 133mm)	14⁷/₈" (378mm)	3³/₈" (86mm)
Grizzly G0619	21⁵/₈" x 6¼" (549 x 159mm)	15⁷/₈" x 5¾" (403 x 146mm)	14⁷/₈" (378mm)	2¾" (70mm)
Grizzly G1005Z	23" x 7½" (584 x 191mm)	12" x 6" (305 x 152mm)	- -	3⁵/₈" (92mm)
Jet JMD-15	23" x 7½" (584 x 191mm)	14" x 6" (356 x 152mm)	- -	3½" (89mm)
Harbor Freight 40939	26" x 6½" (660 x 165mm)	15½" x 6" (394 x 152mm)	7²/₃" (186mm)	2¾" (70mm)

Table 4a. *The range of Far-East-manufactured mills sold in the US, in order of table size.*

Vertical-milling machines in the US
The **Harbor Freight 47158 Micro Mill/ Drill** is the smallest machine imported into

Figure 24. *The Grizzly G8689 Mini Milling Machine.*

the US. It has two variable-speed ranges that provide the ability to drill and end-mill up to ²⁵/₆₄" (9.9mm) and to face-mill ¾" (19mm). It may be used with steel, brass, aluminum, plastic and wood.

The **Grizzly G8689 Mini Mill** has two variable-speed ranges and is supplied with ³/₈" (10mm) and ½" (13mm) collets, a ½" (13mm) drill chuck and a pair of T-nuts.

The **Harbor Freight 44991-8VGA Micro Mill/Drill** has two variable-speed ranges providing the capacity to drill and end-mill ½" (13mm) and face mill 1" (25mm). It comes complete with a drill chuck, draw bar, oil bottle and set of wrenches.

The **Bolton XJ9510 Mill/Drill** has two electronically controlled speed ranges and comes with a ½" (13mm) drill chuck and set of tools.

The **Grizzly G0463 Mill/Drill** has a gearbox giving two variable-speed ranges and a drilling capacity of 1" (25mm). It includes a ⁵/₈" (16mm) drill chuck, two

Spindle to table	Head tilt	Motor	Speeds in rpm	Length x width x height	Weight
7¾"	0°	⅕hp	100 - 1,000	10⅛" x 9⅜" x -	<103lb
(197mm)		(150W)	100 - 2,000	(257 x 238mm x -)	(47kg)
11½"	45°	¾hp	0 - 1,100	20" x 20" x 30¼"	101lb
(292mm)		(560W)	0 - 2,500	(508 x 508 x 768mm)	(46kg)
-	0°	⅘hp	0 - 1,100	- x - x 33¹¹⁄₁₆"	115lb
-		(600W)	0 - 2,500	(- x - x 856mm)	(52kg)
-	45°	½hp	0 - 2,500	-	149lb
-		(375W)		-	(68kg)
14¼"	0°	¾hp	0 - 1,000	27" x 30" x 30¾"	353lb
(362mm)		(560W)	0 - 2,000	(686 x 762 x 781mm)	(148kg)
14¾"	L 90°	1hp	100 - 1,750	27" x 30" x 33¾"	364lb
(375mm)	R 30°	(750W)		(686 x 762 x 857mm)	(165kg)
14³⁄₈"	0°	1hp	110 - 2,580	36½" x 34½" x 40"	374lb
(365mm)		(750W)		(927 x 876 x 1,016mm)	(170kg)
15"	0°	1hp	110 - 2,580	36½" x 37½" x 35½"	440lb
(381mm)		(750W)		(927 x 952 x 902mm)	(200kg)
-	90°	1hp	240 - 2,885	22" x 16" x -	-
-		(750W)		(559 x 406mm x -)	-

T-nuts, two open-end combo wrenches, four Allen wrenches and an extra drive belt.

The very similar **Grizzly G0619 6" x 21" Mill/Drill** has a multi-function digital scale on the quill, a quick-reverse tapping feature, push-button variable-speed control and a digital speed display. It includes a ⅝" (16mm) drill chuck, two T-nuts, a set of wrenches and a spare drive belt.

The **Grizzly G1005Z Mill/Drill** provides twelve spindle speeds and a 1" (25mm) drilling capacity. It has a 3¾" (95mm) round column allowing the head to swivel through 360°. It is supplied with a ½" (13mm) drill chuck, small angle-drilling vise and a carbide-tipped fly cutter.

The **Jet JMD-15 Milling Drilling Machine** has a column diameter of 3⅝" (92mm), twelve belt-driven spindle speeds and the head swivels 360°. It includes a ½" (13mm) drill chuck, a carbide face mill, an angle vise and work lamp. There is an optional stand, collet set and power feed.

The **Harbor Freight 40939-5VGA Vertical Mill** is a turret mill with a 220V 60Hz motor that provides nine spindle speeds. It includes a gooseneck lamp. It can do conventional milling, compound-angle milling, drilling and jig boring. The vertical-milling head can rotate 120°.

Figure 25. *Left, the Harbor Freight 47991 and right, the Grizzly G1005Z mills. Photos courtesy Harbor Freight and Grizzly Industrial Inc.*

Mill	Table size	Table movement	Vertical travel	Quill travel
Grizzly G3102	26" x 6¹/₈"	15⁵/₈" x 6"	14"	3"
	(660 x 156mm)	(397 x 152mm)	(356mm)	(76mm)
Grizzly G3103	26" x 6¹/₈"	15⁵/₈" x 6"	14"	3"
	(660 x 156mm)	(397 x 152mm)	(356mm)	(76mm)
LatheMaster LMT25L	27½" x 7"	19½" x 6⁵/₈"	-	-
	(699 x 179mm)	(495 x 168mm)	-	-
Harbor Freight 42827	27½" x 8¼"	8¼" x 6¹³/₁₆"	8⁵/₈"	4½"
	(699 x 227mm)	(227 x 173mm)	(219mm)	(124mm)
Birmingham RF31	28¾" x 8¼"	18½" x 6¼"	-	5¹/₈"
	(730 x 210mm)	(470 x 159mm)	-	(130mm)
Harbor Freight 33686	28¾" x 8¼"	19¾" x 7³/₈"	5"	5"
	(730 x 210mm)	(502 x 187mm)	(130mm)	(127mm)
Grizzly G3358	28¾" x 8¼"	19¾" x 7"	12"	5"
	(730 x 210mm)	(502 x 178mm)	(305mm)	(127mm)
Birmingham RF40	28¾" x 8¼"	19¾" x 9"	-	5¹/₈"
	(730 x 210mm)	(502 x 229mm)	-	(130mm)
Grizzly G0678	30" x 8"	18" x 7¾"	17¾"	3½"
	(762 x 203mm)	(457 x 197mm)	(451mm)	(89mm)

Table 4b. *The range of Far-East-manufactured mills sold in the US, in order of table size.*

The **Grizzly G3102 Vertical Mill** has nine spindle speeds and the titling head can swivel 360°. It also features a knee, one-shot lubrication and a work light.

The **Grizzly G3103 Vertical Mill with Power Feed** is virtually identical to the **Grizzly G3102** but is fitted with power feed and is somewhat heavier. However, the head may only rotate through 90°. It also has a knee and one-shot lubrication.

The **LatheMaster LMT25L Benchtop Mill** has a drill capacity of 1" (25mm), a face-mill capacity of 2½" (64mm) and an end-mill capacity of ¾" (19mm). It has a digital readout for spindle speed and quill down feed. The spindle is gear driven by a reversible motor. A drill chuck, an arbor and a tool set are included. Options include a stand and power table feed.

The **Harbor Freight 42827-3VGA Geared Head Mill/Drill** has a reversible motor giving six easily changed geared speeds that provide drilling and end-mill

capacities of 1" (25mm) and a face-milling capacity of 2" (51mm). An optional floor stand is available.

The **Birmingham RF31** is available as either a single- or a three-phase mill giving twelve speeds. It has a 4½" (114mm) round column allowing the head to swivel 360°. It can drill up to 1½" (38mm), end mill up to ¾" (19mm) and face mill up to 4" (102mm). Optional variants include an inverter providing variable speeds from 30 - 3,000rpm and a power down feed. An optional floor stand is also available.

The **Harbor Freight 33686-3VGA 1½hp Heavy Duty Mill/Drill** has a 4½" (114mm) column, twelve spindle speeds and a drilling capacity of 1¼" (32mm). Its face-milling capacity is 3" (76mm) and its end-milling capacity is ¾" (19mm). Free accessories include an adjustable carbide-tipped face mill, a 3½" (89mm) angle vise and a ½" (13mm) precision drill chuck with key and arbor.

Spindle to table	Head tilt	Motor	Speeds in rpm	Length x width x height	Weight
12½" (318mm)	45°	1½hp (1.1kW)	240 - 2,760	45½" x 41" x 66⅝" (1,155 x 1,041 x 1,692mm)	800lb (364kg)
12½" (318mm)	45°	1½hp (1.1kW)	240 - 2,760	45½" x 41" x 66⅝" (1,156 x 1,041 x 1,692mm)	900lb (409kg)
13" (330mm)	90°	1hp (750W)	50 - 2,250	38" x 22" x 36½" (965 x 559 x 927mm)	250lb (114kg)
10" (254mm)	45°	1½hp (1,500W)	95 - 1,500	23½" x 28¼" x 30⅞" (597 x 718 x 784mm)	325lb (148kg)
18" (457mm)	0°	2hp (1.5kW)	120 - 2,500	43½" x 38" x 36" (1,105 x 965 x 914mm)	630lb (286kg)
18" (457mm)	360°	1½hp (1.1kW)	120 - 2,500	42½" x 39¾" x 43½" (1,080 x 1,010 x 1,100mm)	<706lb (321kg)
17¾" (451mm)	0°	2hp (1.5kW)	140 - 2,570	41½" x 41½" x 43" (1,054 x 1,054 x 1,092mm)	540lb (245kg)
18½" (470mm)	L 90° R 45°	1hp (750W)	60 - 1,500	43½" x 31½" x 43½" (1,105 x 800 x 1,105mm)	650lb (295kg)
20" (508mm)	90°	1½hp (1.1kW)	200 - 2,250	40½" x 42¾" x 67" (1,029 x 1,086 x 1,702mm)	991lb (450kg)

The **Grizzly G3358 Mill/Drill** provides twelve different spindle speeds. It has a 4½" (114mm) round column allowing the head to swivel through 360° but not to tilt. It is an imperial machine and has a ½" (13mm) Jacobs chuck. It has a drilling capacity of 1¼" (32mm).

The **Birmingham RF40** is available as either a single- or a three-phase system providing six speeds. It has a 4½" (114mm) round column allowing the head to swivel 360°. It can drill up to 1¼" (32mm), end mill up to 1¼" (32mm) and face mill up to 4" (102mm). Options include a power down feed and a floor stand.

The **Grizzly G0678 Variable Speed Vertical Mill** is a knee mill that requires a 220volt single-phase power supply to its three-phase motor to give infinitely variable speeds with a digital readout. The turret swivels 360° and the mill features one-shot lubrication. It comes with a built-in floor stand and a work light.

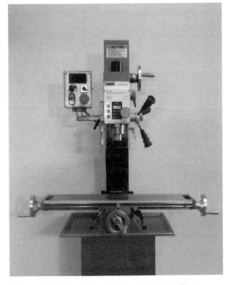

Figure 26. *The LatheMaster LMT25L Photo courtesy LatheMaster Metalworking Tools.*

Mill	Table size	Table movement	Vertical travel	Quill travel
Grizzly G3616	31½" x 9½"	15½" x 8½"	13"	5"
	(800 x 241mm)	(394 x 216mm)	(330mm)	(127mm)
Bolton ZX45	31½" x 9½"	22" x 9"	12"	5"
	(800 x 241mm)	(559 x 229mm)	(305mm)	(127mm)
LatheMaster ZAY7045FG	31½" x 9½"	21¼" x 8½"	-	5"
	(800 x 241mm)	(540 x 216mm)	-	(127mm)
Jet JMD-18	31¾" x 9½"	20½" x 7"	-	5"
	(806 x 241mm)	(521 x 178mm)	-	(127mm)
Jet JMD-18PFN	31¾" x 9½"	20½" x 7"	-	5"
	(806 x 241mm)	(521 x 178mm)	-	(127mm)
Grizzly G1006	32" x 9½"	23½" x 7"	5¼"	5"
	(813 x 241mm)	(597 x 178mm)	(133mm)	(127mm)
Grizzly G1007	32" x 9½"	23½" x 7"	5¼"	5"
	(813 x 241mm)	(597 x 178mm)	(133mm)	(127mm)
Grizzly G1126	32" x 9½"	20½" x 9"	5¼"	5"
	(813 x 241mm)	(521 x 229mm)	(133mm)	(127mm)
Grizzly G0484	32¼" x 9³/₈"	21⁵/₈" x 7⁷/₈"	16½"	4¾"
	(819 x 238mm)	(551 x 200mm)	(419mm)	(121mm)
Birmingham RF45	32¼" x 9½"	21⁵/₈" x 7⁷/₈"	15¾"	5¹/₈"
	(819 x 241mm)	(551 x 200mm)	(400mm)	(130mm)

Table 4c. *The range of Far-East-manufactured mills sold in the US, in order of table size.*

The **Grizzly G3616 Vertical Mill** has a 220V motor giving nine speeds. It is a knee mill with a coolant system, a work light and longitudinal power feed. It includes a drill chuck, a milling vise and a set of collets.

The **Bolton ZX45** is a single-phase mill with six spindle speeds. A set of tools is included with options of a collet chuck, a machine vise, power table feed and a cabinet stand.

The **LatheMaster ZAY7045FG Heavy Duty Dovetail Column Benchtop Mill** has a drill and end-mill capacity of 1¼" (32mm); the face-mill capacity is 3¼" (83mm). The reversible motor can work from 110V or 220V. A drill chuck, a drawbar and a tool set are included. The mill can be fitted with an optional X-axis table feed.

The **Jet JMD-18** and **JMD-18PFN** have a drilling capacity of 1¼" (32mm), can face mill 3" (76mm), end mill ¾" (19mm) and have a column diameter of 4½"(114mm). The **JMD-18PFN** also has a power down feed with manual fine feed. Both mills have twelve belt-driven spindle speeds and a head that swivels 360°. Both include a ½" (13mm) drill chuck, an adjustable carbide face mill, an angle vise and a work lamp. There is an optional stand, a clamping kit, a collet set and a table power-feed add-on.

The **Grizzly G1006 2HP Mill/Drill** can run at twelve different spindle speeds. It has a 4½" (114mm) round column so that the head can swivel 360°. It can handle end mills of ¾" (19mm) and face mills of 3" (76mm). It includes a ½" (13mm) drill chuck, a 3" (76mm) angle vise and a 3" (76mm) face mill.

The **Grizzly G1007 Mill/Drill** differs only from the **Grizzly G1006 Mill/Drill** in that it has a variable-speed longitudinal power feed. It also comes with a ½" (13mm) drill

Spindle to table	Head tilt	Motor	Speeds in rpm	Length x width x height	Weight
15" (381mm)	0°	2hp (1.5kW)	270 - 2,950	41⁵/₁₆" x 41¾" x 80¹⁵/₁₆" (1,049 x 1,060 x 2,056mm)	1,322lb (601kg)
18" (457mm)	90°	2hp (1.5kW)	95 - 1,500	45¹/₃" x 32½" x 41¾" (1,151 x 826 x 1,060mm)	706lb (321kg)
17" (432mm)	90°	2hp (1.5kW)	110 - 1,950	46" x 33" x 42" (1,168 x 838 x 1,067mm)	730lb (332kg)
18" (457mm)	0°	2hp (1.5kW)	150 - 3,000	42½" x 39¾" x 43½" (1,080 x 1,010 x 1,105mm)	660lb (300kg)
26" (660mm)	0°	2hp (1.5kW)	150 - 3,000	42½" x 39¾" x 51½" (1,080 x 1,010 x 1,308mm)	700lb (318kg)
18" (457mm)	0°	2hp (1.5kW)	150 - 3,000	50" x 40" x 48½" (1,270 x 1,016 x 1,232mm)	566lb (257kg)
18" (457mm)	0°	2hp (1.5kW)	150 - 3,000	50" x 40" x 48½" (1,270 x 1,016 x 1,232mm)	566lb (257kg)
16" (406mm)	L 90° R 30°	1½hp (1.1kW)	65 - 1.550	47" x 32" x 55" (1,194 x 813 x 1,397mm)	631lb (287kg)
15½" (394mm)	90°	1½hp (1.1kW)	60 - 1,512	43¼" x 33" x 68" (1,099 x 838 x 1,727mm)	838lb (381kg)
18" (457mm)	L 90° R 45°	1½hp (1.1kW)	60 - 1,500	43½" x 38" x 43" (1,105 x 965 x 1,092mm)	800lb (364kg)

chuck, a 3" (76mm) angle vise and a 3" (76mm) face mill.

The **Grizzly G1126 Gear-Head Mill/Drill** has six spindle speeds. Its 4½" (114mm) round column allows the head to swivel 360°. It can drill holes up to 1" (25mm), can face mill 3" (76mm) and has an end-mill capacity of ¾" (19mm). It comes complete with a variable-speed longitudinal power-table feed.

The **Grizzly G0484 Mill/Drill with stand** has a 220volt motor giving six speeds, a motorized headstock lift system and a longitudinal power feed. It comes with a cast-iron stand and a work lamp.

The **Birmingham RF45** is available as either a single- or a three-phase system providing six speeds. It can drill up to 1½" (38mm), end mill up to ¾" (19mm) and face mill up to 4" (102mm). Options include a power down feed and a floor stand.

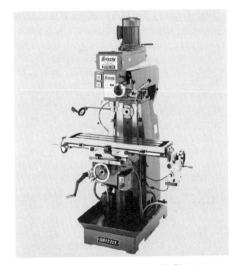

Figure 27. *The Grizzly G3616 Mill. Photo courtesy Grizzly Industrial Inc.*

Mill	Table size	Table movement	Vertical travel	Quill travel
Grizzly G0519	32¼" x 9½"	20½" x 8¼"	15¾"	4¾"
	(521 x 241mm)	(521 x 200mm)	(400mm)	(121mm)
Grizzly G6760	35¾" x 7⁷/₈"	20½" x 9½"	13½"	5¼"
	(908 x 200mm)	(521 x 241mm)	(343mm)	(133mm)
Grizzly G0667X	48" x 9"	29" x 12"	16"	5"
	(1,219 x 229mm)	(737 x 305mm)	(406mm)	(127mm)

Table 4d. *The range of Far-East-manufactured mills sold in the US, in order of table size.*

The **Grizzly G0519 Mill/Drill/Tapping Machine** has a reversible 220volt single-phase motor giving six spindle speeds. It has a ⁵/₈" (16mm) drill chuck and can drill 1¼" (32mm). It also can face mill 3" (76mm) and tap ½" (13mm).

The **Grizzly G6760 Vertical Milling Machine w/Power Feed** has a 220volt motor providing five spindle speeds. It is a knee machine and the head can swivel through 360°. It has a longitudinal power table feed and one-shot lubrication. It has an end milling capacity of ¾" (19mm) and a face milling capacity of 3" (76mm).

The **Grizzly G0667X High Precision Variable Speed Vertical Mill** is a turret

mill that is fitted with a built-in 220volt single- to three-phase inverter driving an infinitely variable speed range motor. the mill has one-shot table lubrication, power down feed, longitudinal power feed to the table and a built-in coolant system.

Horizontal-milling machines

The use of horizontal-milling machines is far less common that vertical mills. As already stated, this is because their key advantage is their ability to allow high metal-removal rates, particularly on parts that are identical, making them more suited to volume production runs than to prototyping. Thus, those who require a single-function horizontal machine that is suitable for a home workshop do not have a great choice.

Typical tools that can be mounted on a horizontal mill include slab mills and face and side cutters as well as slitting saws.

CST
The CST horizontal mills are made in Italy. Their overseas supply arrangements are not published. They offer two variants of the **Zeromatic** as well as the **Zero** machine and the **L.5** and **L.4**. In addition, they can also supply the **L.2** and **L.1** models that have reduced dimensions. Their early milling machines were sold in the UK under the Astra banner.

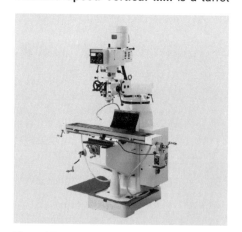

Figure 28. *The Grizzly G0667X mill. Photo courtesy Grizzly Industrial Inc.*

Spindle to table	Head tilt	Motor	Speeds in rpm	Length x width x height	Weight
20" (508mm)	90°	1hp (750W)	120 - 1,970	31½" x 46½" x 57" (800 x 1,181 x 1,448mm)	620lb (282kg)
13¾" (349mm)	90°	2hp (1.5kW)	255 - 1,800	42" x 52" x 75" (1,067 x 1,321 x 1,905mm)	1,620lb (736kg)
18" (457mm)	45°	3hp (2.2kW)	60 - 5,000	57" x 71" x 87½" (1,448 x 1,803 x 2,223mm)	2,156lb (980kg)

The two **Zeromatic** machines are large machines and have power table feed in both directions and manual vertical-table movement. The **Zero** table has manual movement in all three dimensions, as have the smaller **L.5** and **L.4** mills. All the L models have the motor drive contained within the body casting and a table with a ratchet-operated rise and fall. The **L.5** has a two-speed motor, the **L.4** model has a single-speed electric motor; both have two pulleys. The **L.2** table measures 330 x 111mm (13" x 4³/₈"). It has four spindle speeds from 620 - 1,850rpm. CST also offers optional vertical heads including a universal motor-driven one.

Figure 29. A CST (Astra) L4 horizontal mill. Photo courtesy Home and Office Machinery.

	Zeromatic E - C	Zeromatic 2E - 2C	Zero	L.4	L.5
Spindle rpm	180 - 1,600	50 - 1,500	180 - 1,600	410 -1,200	300 - 1,800
Speeds	8	9	8	4	8
Table size mm	800 x 180 (31½" x 7")	800 x 180 (31½" x 7")	710 x 170 (28" x 6¾")	420 x 120 (16½" x 4¾")	500 x 130 (19²/₃" x 5¹/₈")
Spindle to table maximum	320mm (12½")	450mm (17¾")	335mm (13¹/₈")	350mm (13¾")	280mm (11")
Table moves mm	470 x 170 (18½" x 6²/₃")	470 x 170 (18½" x 6²/₃")	420 x 170 (16½" x 6²/₃")	300 x 110 (11¾" x 4¹/₃")	320 x 140 (12½" x5½")
Vertical movement/stroke	310mm (12¼")	450mm (17¾")	355mm (14")	390mm (15¹/₃")	300mm (11¾")
Motor	1.1kW/750W (1½/1hp)	1.5kW (2hp)	1.1kW/750W (1½/1hp)	550W (¾hp)	600/750W (⁴/₅/1hp)
Approx. weight	520kg (1,144lb)	800kg (1,760lb)	450kg (990lb)	200kg (440lb)	220kg (440lb)

Table 5. The details of five CST horizontal-milling machines.

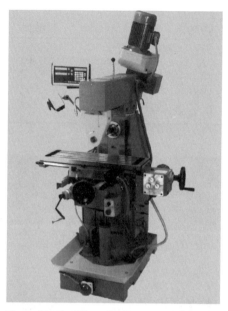

Figure 30. *The Warco Horizontal/Vertical Milling Machine with optional digital readout, set up for horizontal milling. Photo courtesy Warco.*

Combined horizontal/vertical mills

The combination of facilities to carry out both vertical and horizontal milling on the same machine gives advantages in terms of flexibility and space required but, with a single exception, adds to the size and complexity of the machine. The one case is the vertical-milling machine where the head can rotate through ninety degrees, thus providing a simple but very-limited horizontal-milling capability.

The **Opti WF 20 Vario** is a multi-function milling machine for both horizontal and vertical work. It has a single 1.5kW (2hp) motor in a re-positionable head that can be located either for vertical or horizontal milling. When used for vertical milling the head may be tilted up to 90° either way. The table size is 430 x 150mm (17" x 6") in the vertical mode and 485 x 150mm

(19" x 6") in the horizontal mode. The table travel is 310 x 175mm (12¼" x 6⁷/₈"), vertical travel is 290mm (11³/₈") and quill travel is 53mm (2"). The mill measures 570 x 700 x 1,790mm (22½" x 27½" x 70½") and weighs 270kg (594lb).

The **Chester Model T Universal Milling Machine** allows vertical and horizontal milling. Its table measures 800 x 240mm (31½" x 9½"), has a longitudinal travel of 435mm (17¹/₈") and a cross travel of 280mm (11"). Vertical travel is 860mm (33¹/₈") and the head tilts 45° either way. The maximum distance of the vertical spindle from the table is 670mm (26¹/₃") and of the horizontal spindle is 370mm (14½"). The quill travel is 130mm (5¹/₈"). The vertical motor is rated at 2hp (1.5kW) and the horizontal one at 1½hp (1.1kW). These provide a variable vertical-speed range of 50 - 3,000rpm and a variable horizontal one of 20 - 2,500rpm. The mill measures 1,250 x 1,450 x 2,200mm (49¼" x 57" x 86²/₃") and weighs 850kg (1,870lb). The mill comes complete with a three-axis digital readout, a gearbox-driven power feed to move the table on its longitudinal axis, a coolant tray, a work light and some horizontal arbors.

The **Warco Variable Speed Horizontal/ Vertical Milling Machine** shown in Figure 30 is very similar to the **Chester Model T Universal Mill**. Both have two spindles and two motors giving identical power levels and speed ranges. The table size and movement and power feed are the same as is the head tilt, distance from the vertical spindle to the table, spindle travel and the height of the horizontal spindle. The dimensions are virtually identical though the Warco machine is lighter at 750kg (1,650lb). It comes with most of the same items but a two-axis digital readout and a foot stop switch are option. It also has a heavy-duty swivel-base machine vice and a collet chuck with collets.

Figure 32. *A Chester Cestrian set up for horizontal milling. Photo courtesy Chester.*

Figure 31. *The Golmatic MD23 configured for horizontal milling. Photo Pro Machine Tools Ltd.*

The **Grizzly G3617 Horizontal/Vertical Mill** and the **Shop Fox M1009 Vert/Horiz Mill** are similar machines but weigh 2,429lb (1,104kg).

There are three other machines. The **Chester Cestrian Multi-function Machine** can be configured for horizontal milling. It has already been described in Chapter 2 on page 41. The **Golmatic MD23** and **MD24** are other multi-function machines that can be set up to mill horizontally. Their details are in Chapter 2 on page 42.

Second-hand milling machines

There are many quality second-hand milling machines on the market. This process has been hastened by the move to CNC machines by industry. Most have much good life left in them though often they benefit from a major refurbishment.

Astra
There are numerous **Astra L.5**, **L.4** and **L.2** horizontal mills on the second-hand market. These machines, described on page 72, are still in production with CST in Italy. Therefore spares should be readily available from the factory.

Centec
While Centec made a variety of milling machines, the **Mk 2A** and **2B** horizontal-milling machines are similar, desirable but relatively rare. Both have a column-mounted gearbox driven by a 0.5 - 0.75hp ($^2/_3$ - 1kW) three-phase motor that gives six spindle speeds from 85 - 1,400rpm or twice those speeds with a 2,800rpm motor (occasionally with a two-speed 2,800rpm motor giving twelve speeds). The **Centec 2A** has a table of 16" x 4¼" (406mm x 108mm) a longitudinal travel of 9" (229mm), a cross traverse of 4½" (114mm) and a vertical travel of 6" (152mm). The maximum distance from the spindle to table is 6⁵/₈" (168mm). It weighs 360lb (200kg).

The **Centec 2B** has a larger 25" x 5" (635mm x 127mm) table with 14" (356mm) longitudinal, 5" (127mm) cross and 9½" (241mm) vertical travel. The maximum distance from the spindle to table is 10½" (267mm). It also features an improved location for the knee-elevation hand-wheel and has a sturdier build standard. It weighs 500lb (278kg).

The vertical head is a popular Centec accessory that adds vertical milling to the existing machines. Three versions exist; the Mk1 head that has an exposed V-belt

Figure 33. *A Centec 2A horizontal-milling machine. Photo courtesy lathes.co.uk*

Figure 34. *A second-hand Elliott Omnimill 00 combined horizontal/vertical milling machine. Photo courtesy Home and Workshop Machinery.*

in the drive to the spindle, the Mk2 with an enclosed drive and the Mk3, which is the only variant with a quill feed for drilling.

Elliott

Many different Elliott and Elliott/Victoria mills are still circulating in the second-hand market. Those that weigh more than 1,000kg (2,200lb) are not described.

The smallest Elliott miller is the **Model 181** horizontal mill, also known as the **Juniormil 181**. It is very similar to the **Omnimil 00**, or **Junior Omnimil**, that also has a sliding, swivelling and rotating vertical head. It can be fitted with range of various different ¾hp - 2hp (550W - 1.5kW) three-phase, cabinet-mounted motors driving V-belts and pulleys to a two-speed gearbox. It has eight horizontal spindle speeds from 100 - 1,700rpm or 65 - 1,125rpm depending on the motor. A 2¼" (57mm) diameter bar supports the 1" (25mm) diameter arbor, which can be up to 14" (356mm) from the table.

The head of the **Omnimill** has a self-contained V-belt drive unit powered by a ¾hp (550W) three-phase motor that gives seven belt-driven speeds from 200 - 3,600rpm (4,360 in America). The tooling is exchangeable between horizontal and vertical milling on early machines but not on later ones. The 28" x 7" (711mm x 178mm) table with three T-slots is power fed and can move 18" x 6¼" (457mm x 159mm). A coolant pump is built-in.

Both mills measure 38¼" x 38¾" (970mm x 985mm), the **Omnimil 00** is 69" (1,752mm) high and weighs 1624lb (738kg) while the **Model 181** is 56¾" (1,440mm) high and weighs a little less.

Emco Maier

The **Emco Maier FB2** vertical-milling machine has a head that can swivel through 360°. A 250W (⅓hp) motor gives six gear-selected spindle speeds from 120 - 2,000rpm. The maximum height between table and spindle nose is 370mm (14½"). The table measures 630 x 150mm (24¾" x 6"), can move 380 x 140mm (15" x 5½") and the head has a vertical travel of 370 mm (14½"). The machine measures 910mm x 680mm (35⅞" x 26¾") and is 1,730mm (68") high.

Myford
The **Myford VM-B** vertical-milling machine is illustrated on page 43. Its table can be moved 360 x 146mm (14⅛" x 5¾") and its vertical travel is 210mm (8¼"). It weighs around 140kg (308kg). The **VM-F** is a knee-type mill with a power-operated table measuring 760 x 180mm (30" x 7⅛"). A two-speed, 1.1kW (1½hp) three-phase motor gives ten spindle speeds from 160 - 2,840rpm. The milling head has a 174mm (6⅞") stroke and can swivel 45° either way. Myford still offer a wide range of accessories for the **VM-B** and **VM-F**.

Tom Senior
Tom Senior used to make both vertical- and horizontal-milling machines and a number of them are still sold second-hand.

The **S Type** vertical-milling machine has a swivelling head with four spindle speeds from 500 - 3,000rpm driven by a ½hp (375W) single or three-phase motor. The table measures 36" or 28" x 6¼" (914 or 711 x 160mm). Travel is 17" x 5½" (432 x 140mm) and vertical travel is 16¼" (412mm). The longer table increases the lateral movement to 25" (635mm). The machine can hold cutters from ¹/₁₆" - 1" (1.6 - 25.4mm).

The smaller **E Type vertical miller** has a cylindrical column 3" (76mm) across and has a table that is 25" x 4¾" (635 x 122mm). The table travel is 15" (381mm) sideways, 6" (152mm) back and forth and

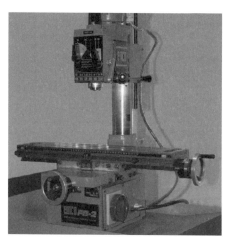

Figure 35. *An Emco Maier FB2 vertical-milling machine.*

14½" (368mm) vertically. The ½hp (375W) three-phase motor gives four spindle speeds from 260 - 2,800rpm and the head swivels 90° either side. The machine measures 27½" x 21½" (699 x 546mm) and is 71" (1,803mm) high. It weighs 360lb (164kg).

The **Junior horizontal milling machine** occasionally comes on the second-hand market. The power-driven table measures 20" x 4¾" (508 x 121mm) and travels 10" x 4½" (254 x 114mm). Vertical movement is 10" (254mm). Depending on the motor, there are twelve speeds from 60 - 4,000rpm or 50 - 2,800rpm. The mill measures 29" x 22" x 60" (737 x 559 x 1,524mm).

The **M1 horizontal milling machine** is powered by a 1hp (750W) three-phase motor that gives twelve spindle speeds from 50 - 1,660rpm. Its table measures 28" x 6¼" (711mm x 160mm), it travels 28" x 9" (394mm x 140mm) and it moves vertically 13" (356mm). It measures 41" x 38" x 63" (1,041 x 965 x 1,600mm) and weighs 1,232lb (560kg).

Figure 36. *A good second-hand example of the Tom Senior S Type vertical-milling machine. Photo courtesy Rondean Machinery.*

The **Major** is in effect a larger version of the **M1**. It is 9" (229mm) wider, 120lb (50kg) heavier with a 2hp (1.5kW) motor. Its table is the same depth but 3" (77mm) longer with a little more movement.

The **VS** was another development of the **Major** with a V-slide over arm rather than the circular bar of the **Major**. It is heavier than both the **Major** and the **M1**.

The **Major ELT horizontal mill** uses a 2hp (1.5kW) three-phase motor to give twelve spindle speeds from 50 - 1,660rpm. It has a table that is 37" x 8½" (940 x 216mm) and moves 28" x 9" (711 x 229mm) with vertical travel of 13" (330mm). It uses the same accessories as the **M1** and **Major** machines.

The **Universal** is quality horizontal mill with a power-operated table that was built in imperial and metric versions. Powered by a 2hp (1.5kW) motor, the spindle can turn at one of twelve speeds from 50 - 1,660rpm. It has a table that is 36" x 8" (914 x 203mm) and moves 20" x 8" (508 x 203mm) with a vertical travel of 12" (305mm). It measures 50" x 57" x 59" (1,270 x 1,448 x 1,499mm). It weighs 1,342lb (610kg).

Chapter 4

CNC machines

For those with the technical interest, and relatively deep pockets, computer numerically controlled (CNC) machine tools give an added depth to model engineering. They enable new and complex parts to be made relatively easily and often without the need for any castings. Instructions from a personal computer (PC), combined with X, Y and Z coordinates, denote the direction, speeds and key points of the shape to be cut.

Machining a part starts with preparing a technical drawing on a PC. Then the data are converted to a form that is suitable for controlling a machine tool. The work to

be done is then, in most cases, simulated graphically before the data are output to control the machining process.

However, to use a CNC machine also requires ownership of a PC that can run suitable software to allow commands to be fed to a computer-controllable lathe or milling machine. For some people, converting a manual machine tool to CNC is a hobby in itself and there are several companies that will supply the necessary bits. However, this chapter concentrates almost entirely on complete lathes and milling machines either with or without a PC.

A number of companies like Boxford, Sherline and Wabeco produce relatively affordable CNC machines. Others like Denford and Emco provide machinery to schools and colleges. Many manufacturers supply industry with professional CNC lathes and mills that are either too large or else too expensive for the average computer-literate model engineer. However, Syil produce a range of machines based on the Far-East-manufactured Sieg series of lathes and mills.

In addition there are thriving companies like Model Engineers Digital Workshop that supply all the various components that are necessary to convert existing machine tools to CNC operation. They offer to

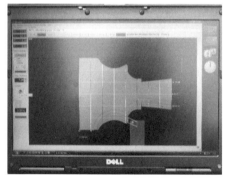

Figure 1. *A Wabeco CNC lathe display shows the tool being applied to a complex shape.*

Figure 2. *The Boxford 160TCL CNC lathe is a compact machine. Photo courtesy Boxford Ltd.*

Figure 3. *The Denford Turn 270 CNC lathe in its enclosure. Photo courtesy Denford Ltd.*

supply comprehensive conversion kits that include a suitable PC and software.

For those who require a fourth axis of movement with a milling machine, there are stepper-motor-driven rotary tables that can be mounted on the mill table. They are described in the next chapter.

CNC lathes

Most CNC lathes suitable for hobby use come from one of three sources. They may be custom-built for the task. They may be relatively small manual machines, with CNC added either by the manufacturer or offered as a kit for installation by the user.

Boxford

The PC-controlled **Boxford 160TCL** lathe comes in an enclosed housing and is unusual (by conventional lathe standards) in having a slant-bed configuration. This allows the swarf to fall into its tray rather than over the tool slides. Both axes have pre-loaded anti-backlash balls screws.

The lathe has a 490W (^{2}/$_3$hp) single-phase motor providing infinitely variable

spindle speeds from 200 - 3,200 rpm. The centre height is 80mm (3^1/$_8$") and saddle and cross travel are 125mm (5"). It has a spindle bore of 20mm (¾") and a tailstock quill travel of 75mm (3"). The machine is 800mm (31½") long, 600mm (23^2/$_3$") wide and 675mm (26½") high with a further 350mm (13¾") clearance needed above the machine. It weighs 130kg (285lb).

The lathe provides simultaneous two-axis operation and each axis is fitted with a DC micro-stepper motor giving 50,000 steps/revolution. It has programmable feed rates of 0 - 1,500mm/min (59"/min). System resolution is 0.005 mm (0.0002").

Using a PC (that is not included), the supplied CAD/CAM (computer-aided design/computer-aided manufacture) software can process drawings to machining routines and it will only allow design of components suitable for lathe work. Drawings from other CAD applications may be imported and processed. 3D simulation of machining can prove the work before it starts. Tool-path graphics are continuously displayed line by line during program writing and machining. Data can be input manually by using the touch-sensitive control panel so the lathe can function without a PC. The basic machine includes the software and cable to connect to a PC,

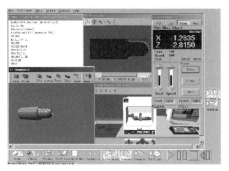

Figure 4. *The VR CNC software used with the Turn 270 CNC lathe. Photo courtesy Denford Ltd.*

instruction and programming manuals, a swarf tray and tools. Accessories include an 80mm (3¹/₈") 3-jaw pneumatic and a 100mm (4") 4-jaw chuck, an automatic 8-station tool post, a coolant system (that requires a compressed-air supply), low-voltage lighting, a workbench and a set of lathe tools.

Denford

Of the CNC lathes made by Denford, one is well-suited to home use. While at first sight, the **Denford Microturn** might seem an attractive proposition, Denford warn that it is not suitable for turning steel, even free-cutting steel.

The **Denford Turn 270** is totally enclosed and its programmable spindle speeds and feed rates make the lathe suited to the task of cutting a range of metals such as steel, free-cutting alloys and aluminium.

Pre-programmed speeds up to 3,500rpm are provided by a 1.1kW (1½hp) motor. The lathe has a centre height of 80mm (3¹/₈") and a distance between centres of 270mm (10²/₃"). The stepper motors give a maximum feed rate of 1,400mm/min (55"/min). Travel along the bed is 225mm (8⁷/₈") and cross travel is 140mm (5½"). The lathe size is 1,000 x 750 x 675mm (39¹/₃" x 29½" x 26½") and it weighs 151kg (332lb).

Figure 5. *The Emco Concept Turn 55 CNC lathe. Photo courtesy Emco Group.*

The lathe includes Windows-based VR CNC turning software. It allows editing and control of CNC files, either offline or when operating the lathe, as well as 2D and 3D graphical simulation of CNC files. A PC is not included but almost any computer that can run Windows XP is suitable. As an option, LatheCAM Designer software can be used to create design work.

The lathe includes a 3-jaw chuck, a quick-change tool post and holder, as well as installation and instruction manuals. Options offered consist of an 8-station programmable turret, a pneumatic chuck, spray-mist coolant and a base stand.

Emco

Emco produce in Europe a range of CNC lathes of which the smallest two are ideal for the home workshop and recommended for hobby use. For both machines, the software can be either FANUC 0T/21T or Siemens 810/840 digitizer, or there is control-keyboard operation.

The **Emco PC TURN 55-II** is a 2-axis CNC enclosed desktop lathe driven by a PC with control software. Powered by a 0.75kW (1hp) single-phase motor, the spindle is lubricated for life and has a variable speed range of 120 - 4,000rpm. The machine has a slant bed with a centre distance of 335mm (13¹/₈"), a centre height of 65mm (2½") and a spindle bore of 16mm (⁵/₈"). The largest work-piece size

Figure 6. *The Optimum CNC controller IV. Photo courtesy Optimum Maschinen.*

Figure 7. *The Proxxon PD 400/CN lathe. Photo courtesy Brimark/Proxxon.*

is 52mm (2") diameter and 215mm (8½") long. The tailstock has a stroke of 35mm (1⅓") and a quill diameter of 22mm (⁷/₈"). The lathe includes a programmable 8-tool turret. The stepper motors move the table 236mm (9¼") by 48mm (1⁷/₈") with an accuracy of 0.008mm (0.0003"). Rapid traverse in both axes is 2m/min (78¾"/min). The machine measures 840 x 695 x 400mm (33" x 22¹/₃" x 15¾") and weighs 85kg (187lb). Total power consumption is 0.85kW (1¹/₈hp).

Options include a PC, a control keyboard, a pneumatic chuck, an electro-mechanical tailstock and a floor stand.

The **Emco Concept Turn 105** is also a desktop machine that is driven by CNC control software, normally WinNC on a PC. Powered by a single-phase 1.9kW (2½hp) motor, the spindle has a variable speed range of 150 - 4,000rpm and a bore of 20.5mm (¾"). It has an 8-station turret tool changer, a pneumatic tailstock, a built-in lubrication system and the main spindle is greased for life. The centre distance is 236mm (9¼") and the centre height is 90mm (3½"). The maximum turning diameter is 75mm (3") and the greatest length is 121mm (4¾"). The saddle and cross slide are driven by stepper motors with a positional accuracy of 0.005mm (0.0002") and a maximum traverse speed of 5m/min (197"/min). Travel is 172 x 55mm (6¾" x 2⅛"). The tailstock quill stroke is 120 mm (4¾") and

its diameter is 35 mm (1⅓"). The machine measures 1,135 x 1,100 x 1,030mm (44²/₃" x 43¹/₃" x 40½") and weighs 350kg (770lb). Options include a PC, a control keypad with TFT display, a coolant system and a machine stand with space for a PC and coolant system.

Optimum

Add-on kits are available from Optimum to convert their **Quantum D210** and **D250**, **Opti D240**, **D280**, **D320 x 630** and **D320 x 920** lathes, described in Chapter 1 pages 20 - 27, to CNC operation. These kits are modular and consist of:

1. Two stepper motors complete with assembly housings, two drive belts, a shaft extension with rotary-shaft seal and screened cable. Some lathes can be fitted with recirculating-ball screws.

2. An electronics box, with monitoring and power supply, to house the control card and three or six motor-control boards.

3. PCTurn software, which suits the needs of most model engineers or the more expensive production-based MegaNC software.

Proxxon

The **Proxxon PD 400/CNC** is based on the manual **PD 400** lathe, described in Chapter 1 on page 15, uses recirculating-ball spindles and a pair of 1.8amp stepper motors to provide saddle and cross slide travel of 300 x 100mm (11¾" x 39¹/₃"). It comes with a CNC control unit, all

Figure 8. *The complete Sherline 8440 CNC package. Photo courtesy Sherline.*

Figure 9. *The Syil C6 lathe with laptop PC and interface box. Photo courtesy Syil.*

connecting cables and CNC software on CD-ROM. It requires a reasonably modern Windows PC that can import CAD files or manually create drawings.

Sherline

The Sherline **Model 4000 and 4400** lathes (see Chapter 1 page 15) are offered as complete CNC packages either with a PC (but no display) or just with the necessary software for installation on the user's PC. Stepper motors each give 200 steps per revolution and the maximum CNC travel speed set at 559mm/min (22"/min).

The **8400 CNC** package includes a **Model 4000** lathe fitted with two stepper motors, a PC (without monitor) with Linux operating system and EMC 4-axis CNC software preloaded as well as a 4-axis driver box inside. All cables, a keyboard and a mouse are included.

The **8440 CNC** package is the same but includes the larger **Model 4400** lathe. Instructions on a CD are included with both systems.

CNC upgrades for existing lathes are also available. An optional rotary table, with stepper motor installed and ready to plug in, can provide a fourth axis.

Syil

Syil UK and Syil America both offer the **Syil C6 CNC** variant of the **Seig C6** lathe described in Chapter 1 on page 22. It has a centre height of 125mm (5") and a 550mm (22") distance between centres. Double-nut ball screws are used and the travel between centres is 495mm (20") driven by a stepper motor. Cross travel is 145mm (5¾"). The CNC version of the machine weighs 200kg (450lb).

The fully enclosed mains-powered CNC control box houses a power supply, up to four micro-step drivers and a parallel PC board controller. Each driver can move its motor through 100 sub-divisions.

While the lathe comes without a PC and its software, the system can readily be connected to any PC running MACH, KCAM4 or Turbo CNC software. Many of the functions are displayed with indicator lights to simplify both setting up and trouble shooting.

Tormach

The **Tormach Duality Lathe** is a normal small lathe with infinitely variable speed, a 4-way tool post and a tailstock. For CNC work, the lathe is mounted to the **PCNC 1100 mill** (see page 95). Both the lathe and mill are available in the US and Canada but purchasers must arrange shipping to other countries. Using the quick-change lathe tool post mounted to

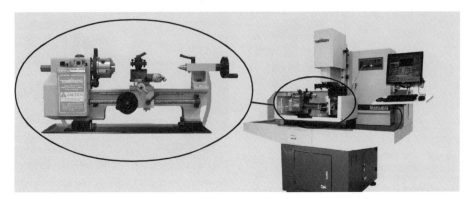

Figure 10. *The Tormach Duality Lathe and the way it fits on the PCNC 1100 mill for CNC use. Photos courtesy Tormach LLC.*

the spindle cartridge of the **PCNC 1100** and CNC lathe software, the motion of the **PCNC 1100** controls the lathe itself.

Thus the mill table moves the **Duality Lathe** instead of using the lathe carriage. Likewise, the cutting tool is rigidly attached to the vertical axis of the mill to replace cross-slide movement. All normal lathe operations such as turning, parting and boring are possible. When complete, the work is immediately ready for milling.

The lathe includes an integral spindle clamp with a 360° index wheel, so that the lathe can be used as a milling fixture.

The lathe swing is 7" (180mm), the distance between centres is 11⁵/₈" (300mm) and the spindle bore is ¾" (20mm). A 250W (¹/₃hp) 115V AC motor

Figure 11. *The Wabeco CC-D6000 E CNC lathe has an excellent capability for rapid work.*

provides two spindle-speed ranges of 50 - 1,100rpm and 100 - 2,500rpm. Lathe dimensions are 33" x 15" x 20" (838 x 381 x 508mm) and it weighs 85lbs (38 kg).

The lathe comes without a mill or PC, but includes a 3-jaw chuck, a dead centre, a quick-change tool post, lathe control software and all necessary cables.

Wabeco
Three different Wabeco lathes (described in Chapter 1 page 16) are also available as the **Wabeco CC-D4000 E CNC, CC-D6000 E CNC** and **CC-D6000 E CNC high speed** lathes. They are supplied with CNC lathe software, a connection cable between the PC and the control system, a factory-installed control unit and feed motors. The CNC lathes can also still be used in a conventional mode with the CNC controller switched off. The feed rate on the **D4000** is from 30 - 500mm/min (1" - 20"/min); on both the **D6000 E CNC** lathes up to 1,000mm/min (40"/min).

The system requirements are not very demanding in computer terms. The nccad software can provide CAD, CAM, CNC, simulation and lathe control. It is used to create the desired component, program

Figure 12. *From left to right, the Boxford 190VMCxi, MidiMILL and 300VMCi CNC mills. All are fully enclosed machines. Photos courtesy Boxford Ltd.*

the lathe for each task and to archive data. The software allows concurrent movement of two axes and can compensate for backlash. Before metal cutting takes place, graphic simulation enables any potential programming errors to be eliminated.

Designing uses the nccad CAD mode or data imported from systems, such as AutoCAD, with automatic conversion to operate the lathe. Instructions can be modified, added or deleted. Values may be input for contours such as the feed, in-feed depth, processing sequence and fine-chip removal. The control panel may be used for direct input of displacement values. Current values are shown on the display.

CNC mills

Milling with CNC appears to be at least as popular as using similar technology with a lathe. It is more flexible as it allows three-axis movement and normally a fourth can be added with a suitable rotary table.

Boxford

All three of the company's CNC mills should fit in a home workshop.

The **Boxford 190VMCxi CNC mill** is an enclosed bench-top computer-controlled vertical-machining centre. It comes with CAM software and 3D graphics and has automatic error checking. It can cut steel, brass, aluminium, plastic and wood

Operated by a PC, the mill has a table measuring 410 x 130mm (16" x 5") with two 10mm ($^3/_8$") T-slots and it moves 225 x 150mm (9" x 6"). Vertical movement is 140mm (5½") and feed rates can be programmed from as little as 10mm/min ($^3/_8$"/min) to the maximum rate of 2m/min (78¾"/min). The resolution of the system is 0.005mm (0.0002").

Infinitely variable and programmable spindle speeds of 350 - 3,500 rpm are provided by a 0.45kw ($^2/_3$hp) DC motor powered from a single-phase supply. The spindle clearance to the table is 202mm (8") and to the column is 130mm (5"). The touch-sensitive control panel incorporates illuminated push buttons and allows the machine to be used without a computer. The mill measures 800 x 210 x 450mm (31½" x 8¼" x 17¾") and weighs 150kg (330lbs).

The range of accessories includes an automatic 6-position tool changer, air mist coolant, a range of work-holding devices and a cabinet base.

The fully enclosed bench-top **Boxford MidiMILL Vertical Machining Centre** is PC-controlled and has a spindle driven by a 320W ($^7/_{16}$hp) DC motor powered from a single-phase AC supply. The motor gives

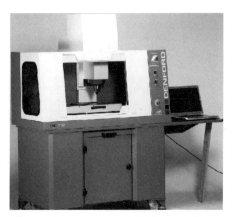

Figure 13. *The Denford VMC 1300 CNC mill with optional bench and PC support (laptop excluded). Photo courtesy Denford Ltd.*

spindle speeds from 200 - 4,000rpm. The table dimensions are 450 x 153mm (17¾" x 6") and it moves 300 x 180mm (11¾" x 7"). Vertical travel is 230mm (9"). The fully programmable feed rate is from 0 - 600mm/min (23$^1/_3$"/min). The machine, excluding computer, measures 840 x 630 x 720mm (33" x 24$^7/_8$" x 28$^1/_3$") and weighs 90kg (200lbs).

The mill includes a touch-sensitive control panel that has illuminated push buttons allowing operation without a PC. It comes with tool holders, a machine light, two slot drills, a ball-end tool and Boxford CAD/CAM Design Tools software. This is an integrated suite of powerful CAD/CAM tools with a variety of drawing tools and manipulation routines to help the user.

The software has routines for CAD and will process the drawings through to a full machining routine. It also allows data imported from other CAD/CAM packages to be machined. It gives 3D simulation of the manufacturing process, including cycle time, allowing work to be proved before machining. Tool-path graphics are continuously displayed on a line-by-line

basis during program writing as well as when machining. Manual data input allows programs to be entered using either line-by-line or conversational programming.

The **Boxford 300VMCi** is a stand-alone PC-controlled vertical-machining centre that provides 3D graphics and includes automatic CAM software. A 1.26kW (1$^2/_3$hp) DC motor powered from single-phase mains electricity gives infinitely variable, programmable spindle speeds from 200 - 4,000rpm. The table measures 505 x 135mm (19$^7/_8$" x 5$^5/_{16}$") and moves 304 x 157mm (12" x 6$^3/_{16}$"). The distance from the spindle to the column is 155mm (6") and to the table varies from 47 - 260mm (1$^7/_8$ to 10$^3/_{16}$"). Vertical movement of the spindle is 213mm (8$^3/_8$"). Pre-loaded anti-backlash ball screws and stepper motors with micro-stepping drives can provide simultaneous movement of all axes with a resolution of 0.01mm (0.0004"). Feed rate can be varied from 0 - 5,000mm/min (197"/min) and system resolution is 0.005mm (0.0002"). The mill measures 1,425 x 820 x 1,550mm (56" x 32¼" x 61") and weighs 500kg (1,100lbs).

The mill has a touch-sensitive control panel with illuminated push buttons for operation without a PC, a 6-position tool storage rack, CAM software with 3D graphical simulation and automatic error checking, and full graphical on-screen test simulation prior to machining. Graphics are continuously displayed line-by-line during program writing and machining. The software can be linked to most CAD/CAM packages.

An assortment of accessories includes an automatic 6- or 8-position tool changer, a range of work-holding devices and a variety of tools and holders.

Denford

While the range of Denford machines is primarily aimed at the educational market,

86

one is very suitable for model engineering use. But the **Micromill CNC mill**, like the **Microturn** lathe, is unsuitable for cutting any steel, even free-cutting steel.

The **VMC 1300 CNC mill** is a 3-axis totally enclosed machine that is either floor standing or bench mounted. The ability to program spindle speeds and feed rates makes it ideal for working many metals such as free-cutting alloys and steel.

The mill is fitted with a 1.1kW (1½hp) motor giving programmable spindle speeds up to 4,000rpm; the **1300 Pro** has a 1.6kW (2hp) motor giving a top speed of 6,000rpm. Stepper motors move the 600 x 180mm (23$^2/_3$" x 7") table at up to 5,000mm/min (197"/min); 4,500mm/min (177"/min) if 3D profiling. X-axis travel is 375mm (14¾") but only 250mm (9$^7/_8$") if an automatic tool changer is used. Y-axis travel is 150mm (6") and the head moves 235mm (9¼").The maximum distance from the head to the table is 305mm (12"). The machine measures 1,300 x 750 x 1,000mm (51$^1/_8$" x 29½" x 39$^1/_3$") and weighs 353kg (777lb).

Both **VMC 1300** variants come without a PC but complete with VR CNC milling software necessary to control the mill. Almost any computer with Windows XP, NT or 2000 and a USB connection will suffice. Also included is QuickCAM 2D Design; a 2D CAD package. The VR CNC milling software will also link to packages such as TechSoft Design Tools - 2D Design, CorelDraw, Pro/DESKTOP, ArtCAM and Autodesk Inventor SolidWorks when used in conjunction with QuickCAM Pro.

The mill comes with a power drawbar with manual actuation, work-holding clamps and manuals. Options include a table-mounted 6-station automatic tool changer, a pneumatic vice, a spray-mist coolant system, a substantial machine bench and a fourth-axis programmable rotary table.

Figure 14. *The Emco Concept Mill 105 CNC milling machine. Photo courtesy Emco Group.*

Emco

The Emco range of CNC milling machines includes two desk-top machines; the **Concept Mill 55** and **105**. They should suit home-workshop use but Emco recommend only the smallest for the hobby market.

The **Emco Concept Mill 55** is a small enclosed desk-top machine equipped with an optional 8-station automatic tool changer. It is powered by a 750W (1hp) motor giving an infinitely variable spindle speed range of 150 - 3,500rpm. The main spindle is greased for life. The table size is 420 x 125mm (16½" x 5") with two 11mm ($^3/_8$") T-slots. The stepper motors provide X, Y and Z movements of 190 x 140 x 260mm (7½" x 5½" x 10¼"). X-axis positional accuracy is 0.006mm (0.0002") and for the other two axes is 0.008mm (0.0003"). Effective Z-axis movement is limited to 120mm (4¾").

The rapid traverse rate on all axes is 2m/min (79"/min). The maximum tool diameter and length are respectively 40mm and 50mm (1½" and 2"); 60mm and 100mm (2$^1/_3$" and 4") without the changer.

The mill measures 960 x 1,000 x 980mm (37¾" x 39$^1/_3$" x 38½") and weighs 190kg (418lb) with the automatic tool changer. The total electrical load is 850W (1$^1/_8$hp) single phase.

Figure 15. *The Microproto Systems DSLS 3000 based on the Taig mill hardware. Photo courtesy Microproto Systems.*

Options include a floor stand with space for a PC, an NC-indexing attachment to give a fourth axis and a run-up spindle providing 14,000rpm.

The **Emco Concept Mill 105** is a portable enclosed desktop machine with a 10-tool automatic changer driven by CNC software running on a standard PC.

The spindle is lubricated for life and is driven by a 1.1kW (1½hp) motor giving an infinitely variable speed range of 150 - 5,000rpm. The maximum tool diameter and length are 55mm (2¹/₈") and 50mm (2"). The milling table, with two 11mm (³/₈") T-slots, has a clamping surface of 420 x 125mm (16½" x 5"). Three stepper

motors provide X, Y and Z movement of 200, 150 and 250mm (7⁷/₈", 6" and 9⁷/₈") but the effective Z stroke is 150mm (6"). Positional accuracy is 0.003mm (0.0001") on the X and Y axes but 0.004mm (0.00015") on the Z one. Rapid traverse on all three axes is 5m/min (197"/min).

The machine measures 1,135 x 1,100 x 1,100mm (44⁵/₈" x 43¹/₃" x 43¹/₃") and weighs 400kg (880lb). Power consumed is 1.4kW (1⁷/₈hp) single phase.

The options include a fourth-axis NC-dividing attachment, a machine stand with PC compartment, a built-in desk for a control keypad and monitor as well as space for a coolant system.

Microproto Systems
The milling hardware used as a basis for the **MicroMill 2000LE** and **DSLS 3000** is the **Taig Micro Mill** described in Chapter 3 page 51. Like Taig, Microproto Systems is also an American company.

The key differences between the two systems lie in the CNC additions. The range includes the original open-loop-based **2000LE** and the newer closed-loop **DSLS 3000**. Both are made in the US and are full desktop machining systems.

These two CNC mills are powered by a ¼hp (0.2kW) electric motor providing a spindle-speed range with six pulley steps of 1,100 - 10,000 rpm. The vertical head movement is 6" (152mm) and the maximum distance from the spindle to the table top is 8" (203mm). An imperial collet chuck with collets is included and the spindle hole is ¹/₃" (8.7mm). The machines are 26³/₈" (670mm) high, 22" (559mm) wide, 21" (533mm) deep and weigh 88 lbs (40kg).

200oz-in (223g-mm) stepper motors are used on all three axes. An amplifier and a port for the fourth-axis are included. They provide 200 steps per revolution giving a 0.000125" (0.003mm) resolution and a mechanical repeatability of 0.0005"

(0.013mm). The size of the electronics case is 10" x 8" x 5" (254 x 203 x 127mm). The software is Mach3 that can run on most PCs with Windows XP. A connecting cable to the mill electronics and a full user manual are both included. Input power requirement is single-phase 110v AC.

System options include MeshCam 3D software that imports dxf, stl and bmp files or BobCad-Cam design and manufacturing software, a fourth axis rotary table and a 3D digitizing probe.

The **MicroMill 2000LE** uses bi-level chopper-drive technology and an open-loop control system to move the head and table. It provides a maximum rate of travel of 30"/min (760mm/min). Head movement is 6" (152mm). The table has 3 T-slots, measures 15¾" x 3½" (400 x 89mm) and its travel is 9" x 5¾" (229 x 146mm).

The **MicroMill DSLS 3000** has Digital Sync Lock Servo (DSLS) control, which employs a servo-control algorithm that locks the encoder feedback signal with the command signal. Thus the system follows the commanded velocity and position. The addition of an optical encoder converts an open-loop stepper-motor system, with step and direction software, to a genuine closed-loop servo system. The system is fitted with high-resolution 1600 counts/revolution encoders. This solution gives higher feed rates and potentially greater accuracy than the open-loop **2000LE**. The top traverse speed is 100"/min (2,500mm/min) while a continuous-contouring feed rate of up to 59"/min (1,500mm/min) is possible. The table, with three T-slots, measures 18³/₈" x 3½" (467 x 89mm) and its travel is 12" x 6" (305 x 152mm).

Model Engineers Digital Workshop
As well as offering the **Taig/Microproto Systems MicroMill 2000LE** and **DSLS 3000**, this British company can provide a fully enclosed **Industria3** version of the

Figure 16. *The Industria3 CNC mill from Model Engineers Digital Workshop. Photo courtesy Model Engineers Digital Workshop.*

MicroMill DSLS 3000 as well. It can also offers various options to upgrade existing **2000LE** mills with closed-loop control. In addition, these mills can be supplied with a more powerful 0.25kW (¹/₃hp) variable-speed motor giving 0 - 10,000rpm or a Kress 800W (1hp) high-speed spindle providing 10 to 30,000rpm.

The heart of the **Industria3** is the **MicroMill DSLS 3000** closed-loop system, which is housed in a totally enclosed cabinet that measures 1,045 x 614mm (41" x 24") and is 735mm (29") high. The mill and cabinet weigh 65kg (143lb). The system still has a separate electronics housing and an optional Mach-in-a-box ITX-based computer with a reduced version of Windows XP and a copy of the Mach3 CNC software.

The **Mill-in-a-box** is a total package consisting of a CNC-ready Taig mill, stepper motors, mounts, a controller, a Mach-in-a-box computer (a small profile ITX-based computer configured with a stripped-down version of Windows XP and a fully licensed copy of the latest Mach3 CNC software) as well as a high-speed spindle already configured and working. A fourth-axis rotary table can be added as an optional extra.

Figure 17. *The Proxxon FF 500 CNC mill. Photo courtesy Brimark/Proxxon.*

Optimum

Optimum offers add-on kits to convert their **Opti BF20, 30, 40** and **46 Vario** mills for CNC operation. They use the same CNC electronic-control boxes as their lathes illustrated in Figure 6 on page 82. These kits comprise several modules:

1. Stepper motors complete with housings, couplers, shaft extensions and screened cable; for some mills there are also recirculating-ball screws.
2. An electronics box with monitoring and power supply that also houses the system-control board and three or six motor-control boards.
3. PCTurn software that is well suited to the needs of model engineers or the more expensive and industry-orientated MegaNC software.

Proxxon

The **Proxxon FF 500 CNC Micro Mill**, apart from recirculating ball spindles and powerful stepper motors, is almost identical to the manual **Micro Miller FF 500** described in Chapter 3 on page 49. The table can move 310 x 100mm (12¼" x 4") and vertical movement is 220mm (8²/₃"). The mill comes with a CNC control unit, connecting cables and a CNC software program to run on a Windows PC.

Figure 18. *The add-on stepper motors for the Opti BF30 Vario. Photo courtesy Optimum Maschinen.*

Rishton

Now owned by HME Technology, the **HME Technology/Rishton CNC M10 Milling Machine** is a totally enclosed bench-top machine that is small enough to fit in a home workshop. An 800W (1hp) DC motor provides a step-less spindle-speed range of 300 - 3,000 rpm with a vertical travel of 160mm (6¼"). The table measures 450 x 130mm (17¾" x 5¹/₈") and can travel 250mm x 130mm (9⁷/₈" x 5¹/₈"). The table and head are moved by 200 steps-per-revolution stepper motors, which are half-stepped to improve accuracy, driving ball-screws to give a system resolution of 0.01mm (0.0004").

Sherline

Sherline offers two CNC mills, the **P/N 8540** that includes a **5400 mill** and the **P/N 8020** that is similar but substitutes the 8-direction **2000 mill**; each offered either in imperial or metric form. Both of the basic mills are described in Chapter 3 pages 49 and 50.

Additions to the base-line mills include three stepper motors with four drivers (the fourth for an optional rotary table), cable and a 24volt power supply installed in the

90

Figure 19. *The CNC M10 bench-top milling machine. Photo courtesy HME Technology Ltd.*

Figure 20. *The Sherline P/N 8540 includes a deluxe Sherline 5400-CNC mill complete with three stepper motors, a PC and software. Photo courtesy Sherline.*

PC box. The maximum CNC positioning speed in all three axes is 22"/min (559mm/min) and feed rates when cutting metal should normally not exceed 6"/min (152mm/min). The CNC stepper motor holding torque is 136oz-in (152g-mm).

The package includes a new PC that has the Linux operating system and EMC 4-axis CNC software pre-installed on its hard drive. There are backup CD's, full instructions, free utilities to translate CAD files into g-code and two free CAD programs, QCad and Synergy. A monitor is not included.

Systems without computers, as well as CNC upgrades for those who already own a Sherline mill, are both available.

Sieg

Both Arc Euro Trade in the UK and Harbor Freight in the US offer two Chinese-built Sieg CNC Hobby Mills. UK-based on-line support is also offered. The specification of the PC needed to run the software is not demanding though at present it will only operate with Windows 2000 or XP. The machines are supplied with a parallel cable (the only connection between PC and mill), Mach3 demo software and custom configuration files with a quick

start, a guide and technical manual for the Mach3 software as well as a toolkit.

The **Sieg KX1 CNC Hobby Mill** is quite a small milling machine that needs to be connected to a Windows PC running the Mach3 software that is provided with the machine. It is fitted with a CNC-controlled 500W (^{2}/$_3$hp) brushless DC motor driving the spindle via a toothed belt to provide variable speeds from 250 - 7,000rpm. The whole machine is mains operated.

The mill includes three NEMA 23 hybrid stepper motors with drivers that directly turn precision ball screws. An additional stepper driver for a fourth axis is pre-wired with an external socket.

The effective table size is 400 x 145mm (15¾" x 5¾") with X- and Y-axis travel of 260mm (10¼") and 115mm (4½"). Vertical head movement is 185mm (13"). The positional accuracy is 0.01mm (0.0004"). The distance from the head to the table can range from 70 - 255mm (2¾" - 10") while the throat is 140mm (5½"). The bed ways are completely covered and lubrication is by press-button oilers.

The mill measures 630 x 630 x 630 mm (24¾" x 24¾" x 24¾") but requires a space

Figure 21. *The Sieg KX1 CNC Hobby Mill. Photo courtesy Euro Arc Trade.*

Figure 22. *A PC running Mach3 software to drive a Sieg KX3 CNC mill.*

910mm (35⁷/₈") wide and 730mm (28¾") high. It weighs 86kg (189lb).

The **Sieg KX3 CNC Hobby Mill** is a full 3-axis stepper-motor-driven machine that allows direct connection to any PC that runs the popular Windows-based Mach3 CNC control software. The mill is fitted with precision ball screws and also has three direct-drive NEMA 34 hybrid stepper motors with drivers to control the three main axes. No hand wheels are fitted. An extra driver for a fourth axis is pre-wired with an external socket.

The mill has a 1kW (1¹/₃hp) closed-loop variable-speed brushless DC motor that is CNC controlled and drives the spindle via a toothed belt. Spindle speeds vary from 200 - 3,500rpm and the whole machine is mains powered.

The mill has a one-shot lubrication system and fully covered bed ways. The effective table size is 470 x 160mm (18½" x 6¹/₃"), with three T-slots. The mill travel in the X-axis is 295mm (11²/₃"), in the

Y-axis is 150mm (6") and in the Z-axis is 275mm (10⁷/₈"). Positional accuracy is 0.01mm (0.0004"). The distance from the head to table ranges from 80 - 355mm (3¹/₈" - 14") and the throat is 190mm (7½"). The mill measures 850 x 900 x 940mm (33½" x 35½" x 37") but requires a space 1,145mm wide (45"). It weighs 201kg (462lb). There is an optional floor stand.

Syil

Both in the UK and the US, Syil import a range of Sieg manual mills that have been converted for CNC operation. These mills differ from the Sieg CNC mills offered by Arc Euro Trade and Harbor Freight.

The **Syil CNC U2 Mini Grinding and Milling Machine** is easily re-configured from vertical to horizontal milling or to surface grinding. The heart of the machine is the **Sieg U2 Mill/Drill/Grinder** that is described in in Chapter 3 page 55. Spindle speeds can vary from 80 - 2,500rpm. Two stepper motors with 100 steps per revolution give the table travel of 230 x 120mm (9" x 4¾"). Vertical head movement is 180mm (7") provided by a more-powerful

Figure 23. *Syil CNC U2 milling machine. Photo courtesy Syil.*

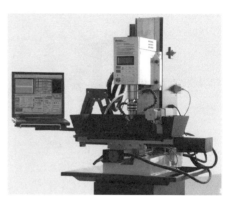

Figure 24. *The Syil X4 Plus CNC mill with user-supplied laptop PC. Photo courtesy Syil.*

stepper motor. All three axes have double-nut precision ball screws, heavy-duty bearings and direct-drive couplers that provide an accuracy of 0.038mm (0.0015") and a repeatability of 0.01mm (0.0004"). The rapid transverse speed is 2.03m/min (80"/min). The system can be operated by software such as Mach or KCAM4. A cable is provided to connect the controller to a user-supplied PC that can be installed on the heavy-duty tray.

A size-4 grinding wheel is provided. A fan-cooled stand houses the control electronics and power supplies, leaving space for an optional coolant system or storing tools. The machine weighs 240kg (529lb).

The **Sieg X2 CNC Mini Mill** described in Chapter 3 page 54 has been converted by Syil to operate as a CNC machine with a stand-alone fan-cooled control box. It contains the CNC interface board, three stepper drivers and a 500W (^{2}/₃hp) power supply. The three axes are updated with precision ball screws, heavy-duty bearings and direct-drive couplers to ensure high accuracy and repeatability.

Three stepper motors move the table 231 x 86mm (9" x 3^1/₃"). Vertical-head movement is 203mm (8"). The motors provide 100 steps per revolution. The system is aimed at use with a PC running software such as Mach, or KCAM4.

For current owners of an **X2 Mini Mill**, a CNC-conversion kit is offered. It consists of a complete set of parts for assembly by the user, including a CNC controller.

The **Syil X4 Plus CNC Milling Machine** employs a 1kW (1^1/₃hp) motor to provide spindle speeds from 100 - 3,300rpm. The maximum distance from the spindle to the table is 380mm (11^7/₈"). The table size is 550 x 160mm (21^2/₃" x 6^1/₃") with a travel of 400 x 160mm (15¾" x 6^1/₃"). Three stepper motors with dual 300W power supplies and with 100 divisions per revolution give a maximum feed rate of 1.5m/min (59"/min). Each axis has ball screws, plain bearings and couplers that will ensure an accuracy of 0.013mm (0.0005") and a repeatability of 0.01mm (0.0004"). The power supplies and control system are in a fan-cooled back shell. A multi-function LCD display and touchpad controls provide manual jogging, axis direction, coolant and spindle control. Computer control is implemented

Figure 25. *The Syil X5 CNC mill. Photo courtesy Syil.*

via a user-supplied PC running Mach3 software that may be fitted on a 3D bracket tray. A floor stand, a coolant system and a fourth axis are options. The machine measures 760 x 720 x 930mm (30" x 28$^1/_3$" x 36$^2/_3$") and weighs 200kg (440lb).

The latest **Syil X4 Speedmaster CNC Milling Machine** is fitted with a 0.75kW (1hp) single-phase motor that provides the water-cooled spindle with a speed range from 8,000 - 24,000rpm. The table size is 550 x 160mm (21$^2/_3$" x 6$^1/_4$") and its travel is 280 x 160mm (11" x 6$^1/_4$"). Vertical head movement is 270mm (10$^2/_3$"). The mill is fitted with double-nut ball screws and has a repeatability of 0.015mm (0.0006"). It comes with a manual lubrication system, a dual-cooling system, a computer arm and a fourth-axis interface pre-installed. The mill is 760mm (30") wide, 700mm (27½") deep and 850mm (33½") high. It

weighs over 250kg (550lb) and there is an optional floor stand.

The **Syil X5 CNC Milling Machine**, known in the US as the **Syil BF20 (X5)**, is the Syil conversion of a popular-sized mill and retains hand wheels on each axis for use in manual mode. The X, Y, and Z axes are updated with ball screws, providing high-positional accuracy. 100 sub-division movement gives performance close to servo-based systems. An optional PC gives full computer control.

A 1kW (1$^1/_3$hp) motor provides spindle speeds of 100 - 3,300rpm. The table measures 500 x 160mm (19¾" x 6$^1/_3$") and its travel is 400 x 160mm (15¾" x 6$^1/_3$"). Rapid traverse speed is 1.5m/min (59"/min). The machine has an accuracy of 0.013mm (0.0005") and a repeatability of 0.01mm (0.0004"). The head travel is 380mm (15") and the machine weighs 70kg (154lbs).

The recent **Syil X5 Plus CNC Milling Machine** and the **Syil X5 Speedmaster CNC Milling Machine** both come with a 0.75kW (1hp) single-phase AC motor. The table is 522 x 160mm (20½" x 6$^1/_4$") and moves 280 x 160mm (11" x 6$^1/_4$"). The vertical head travels 270mm (10$^2/_3$"). The mill has double-nut ball screws and a repeatability of 0.01mm (0.0004"). It comes with a pre-fitted manual lubrication system, a dual-cooling system, a fourth axis interface and a computer arm. The mill measures 920 x 790 x 950mm (36$^1/_4$" x 31" x 37$^1/_3$") and it weighs over 350kg (770lb). The **X5 Plus** has a spindle-speed range from 200 - 5,000rpm while the **X5 Speedmaster** employs a water-cooled spindle with a speed range from 8,000 - 24,000rpm.

The **Syil X6 Plus CNC Milling Machine** has a table that measures 1,000 x 295 mm (39$^1/_3$" x 11$^1/_3$") that can move 600 x 300mm (23$^2/_3$" x 11¾"). Powered by a 1.5kW (2hp) motor, which gives speeds from 200 - 3,500rpm, the maximum distance from the spindle to the table is 320mm (12^2/$_3$"). The control system uses Mach3 software with feed rates up to 3m/min (118"/min). The packed mill is 1,400 x 1,300 x 2,150mm (55" x 51" x 84^2/$_3$") and weighs just over 800kg (1,760lb).

The **Syil X7 Plus CNC Milling Machine** table measures 800 x 250mm (31½" x 9^7/$_8$") and can move 400 x 240mm (15¾" x 9½"). Vertical travel is 400mm (15¾"). The system repeatability is 0.01mm (0.0004"). A 1.5kW (2hp) single-phase motor gives spindle speeds of 200 - 6,000rpm. The mill measures 1,250 x 1,050 x 1,700mm (49¼" x 41^1/$_3$" x 67") and weighs about 850kg (1,870lb). A 3" (79mm) single LCD display screen and a 'one-shot' lubricating-oil system together with a pre-installed fourth-axis interface are provided.

Taig

The two variants of the **Taig CNC desk-top mill** use the **Taig Micro Mill** either with the small or large table and the additions produced by Microproto Systems to configure them as a full CNC mills. They can cut steel to a depth of 3mm (^{1}/$_8$") with a 3mm (^{1}/$_8$") slot mill. Both mills have already been described on pages 51 and 88.

The CNC version of the **Micro Mill** is larger at 22" x 21" (559 x 533mm), taller at 26^3/$_8$" (670mm) and weighs 85lb (39kg). It is fitted with a ¼hp (188W) motor that provide six increased spindle speeds from 1,000 - 10,000 rpm.

Tormach

The **PCNC 1100 Series II** is a bench-top 3-axis CNC mill that is shown in Figure 10 on page 84 and in Figure 26 above. The mill comes with a 1½hp (1.1kW) single-phase motor that gives variable spindle speeds from 100 - 5,100 rpm using a belt drive with two ratios. The mill requires a

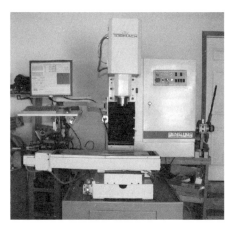

Figure 26. *The Tormach PCNC 1100 mill. Photo courtesy Tormach LLC.*

115V AC single-phase electrical supply. The table size is 34" x 9.5" (867 x 241mm) with 3 T-slots and its travel is 18" x 9½" (457 x 241mm). Head movement is 16¼" (413mm) and the maximum distance from the spindle to the table is 17" (432mm); from the spindle to column it is 11" (279mm). The three axes are moved by stepper motors with micro-stepping drivers and give a maximum feed rate of 65"/min (1.7m/min).

The mill measures 43½" x 46" x 57¼" (1,105 x 1,168 x 1,454mm) but it requires a width of 63¾" (1,619mm) to allow for full table movement. It weighs 1,130lb (514kg).

In the minimum configuration, the mill comes without the optional stand or PC. However, it includes the CNC control software with all functions, including fourth-axis support and interconnecting cables, which allow it to run on a PC with Windows XP or Vista.

Wabeco

Wabeco produce a range of six CNC mills based on the manual machines described

Figure 27. The Wabeco CC-F1210 E CNC milling machine with optional rotary table.

Figure 28. An Emco Compact 5 CNC lathe complete with monitor. Photo lathes.co.uk.

in Chapter 3 on pages 52 and 53. These mills are designated **CC-F1200 E**, **CC-1200 E High Speed**, **CC-F1210 E**, **CC-F1210 E High Speed**, **CC-F1410 LF** and **CC-F1410 LF High Speed**. The motors, speed ranges, table sizes and movements are all the same as the manual machines. All six machines may be programmed for a feed-path rate of up to 1.2m/min (47"/min) and can be used conventionally or with CNC. The connecting cable between the user-supplied PC and the control system of the mill, as well as the nccad milling software, are provided together with a factory-installed control unit with integrated user controls. Both metric and imperial versions can be supplied.

Second-hand machines

While older second-hand CNC machines may well be in excellent condition, their electronic technology has been overtaken by the latest PCs, software and control systems, and programming them can be a lengthy process. However, Boxford will upgrade old **125TCL** machines to current production standards for around one-third of the cost of a new **160TCL**.

Ex-education CNC machines, mainly **Emco Compact 5 CNC** lathes and **Emco F1 CNC** milling machines were originally supplied in the 1980's to a large number of schools and further-education colleges but production was discontinued in 1992. Now quite a few of these machines are becoming available on the second-hand market at prices that the home user can afford! And Emco can still supply the majority of spare parts and accessories and several companies will undertake refurbishment. However, it is important to establish whether the system is first, second or third generation because the control capabilities improve considerably with each generation.

Chapter 5

Lathe and mill tooling/fixtures

All lathes and milling machines, whether manually or CNC operated, will need some cutting tools, fixtures and holding devices as well as measuring equipment. These will help to obtain the optimum flexibility and performance from the machines.

Most Far-East imports are supplied with a range of accessories. Those produced in the industrialised nations usually come with a minimum of add-on items though a lot more are offered as options by the original-equipment manufacturers.

Many specialist suppliers can provide different attachments, fixtures and fittings as well as various cutting tools. Most of these manufacturers will list the names of compatible lathes and milling machines. However, it is essential to ensure that items suit the machine tool on which they will be used. It is important to check whether the size of any chosen ancillary will fit and whether its mounting method is compatible. Tool-shank sizes, thread types and sizes as well a T-nut dimensions must also be taken into account.

As with machines themselves, the quality of ancillary items depends on which company offers them and where they were manufactured. Once again, items produced by reputable companies in the industrialised nation tend to be better than some of those that are made in the developing world. Page 118 gives contact details of a number of companies which supply popular ancillaries.

Many accessories can be built in the home workshop. The modelling press continues to produce books, articles and plans covering a range of useful devices.

Inevitably, some items are suitable for use only on a lathe; others just for work on a milling machine. But most find applications on both types of machine, especially as lathes are often used for relatively simple milling tasks. Items such as chucks, mills and coolant systems are found on both types of machine.

To use a lathe when the work piece is turning, a cutting tool is needed and there is a considerable choice of different types. In addition, if the work is mounted on a faceplate, on the cross slide or on a vertical slide, a means of holding it in place will have to be devised.

Milling machines classically employ many different types of mill to cut work pieces but drills and fly cutters also find application. Sometimes, as in the case of a dividing head, a component may need to be rotated in a chuck between applications of the cutting tool, whether in a lathe or mill.

Figure 1. *A wide range of different steel tools for lathes. From left to right: left- and right-hand turning, straight finishing, corner, parting, boring, threading and internal threading tools.*

Figure 3. *A set of brazed carbide-tipped lathe tools. Photo courtesy Chronos.*

Cutting tools

There are lots of different sizes and shapes of tool that are able to cut metal when used in a lathe or milling machine Many are applicable to both types of machine.

Lathe tools
Lathe tools can be classified in many ways. There are:
Roughing and finishing tools.
Left and right corner tools.
Straight and corner knife tools.
Round-nosed and profiling tools.
Thread-cutting and parting tools.
Some examples are shown in Figures 1 - 3. A selection will be needed by anyone who is a beginner and the range of tools will increase as new work demands different ones. The shank size must be chosen to

Figure 2. *Two indexable carbide-tipped tools.*

fit the lathe's tool post. While lathe tools can readily be made by grinding a length of high-speed steel (HSS), it is common practice to purchase ready-made ones.

Carbide-tipped tools
The introduction of carbide-tipped tools with the tip brazed onto a steel shank has made turning easier and working with harder materials relatively simple. The only real weakness of carbide-tipped tools is their relative vulnerability to chipping, especially when making intermittent cuts. Carbide does, however, require a green stone to be fitted to any grinding machine used to sharpen it.

Replaceable carbide-tipped tools
The introduction of indexable carbide tips allows each side of a triangular or rectangular tip to be rotated as the cutting edge becomes worn. The tips may be indexed or replaced using an Allen or similar key.

Milling cutters
The primary cutting tools that are used for milling are designed either for vertical or for horizontal machines. The majority of vertical-milling cutters are specified by the material, diameter, length, helix angle, type of shank and shank diameter, which must fit the milling chuck. Horizontal mills specify the type, material, width,

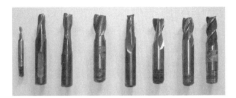

Figure 4. *A selection of HSS slot, end and bull-nose mills for use when vertical milling.*

diameter and hub size. Milling cutters are often used in lathes to carry out milling with a vertical table or angle plate by those who do not own a milling machine. As with lathe tools, mills with carbide inserts are becoming increasingly popular.

End mills
These mills normally have four cutting faces and are supplied as standard length or long-series end mills. They are used for facing, slotting and profile milling. Rough-cut end mills are ideal for rapid removal of metal while face-milling cutters are well suited to taking heavy cuts.

Slot mills
For cutting holes and slots without first drilling a hole or starting at an edge, the slot mill normally has just two cutting edges. It may gently be plunged into metal in a similar way to a drill.

Figure 5. *Three indexable carbide-tipped mills. Photo courtesy RDG Tools.*

Figure 6. *A classic Woodruff-key slot cutter. Photo courtesy RDG Tools.*

Dovetail, T-slot and ball-nose cutters
For those who wish to cut specially shaped slots, a wide range of different cutters are available. A typical example is shown in Figure 6.

Horizontal slab mills
Mills for use on horizontal machines tend to be designed to allow heavy cuts to be taken on large, flat surfaces. A selection is shown in Figure 7.

Side and face cutters
These mills have cutting edges on their periphery and on the sides of their teeth allowing them to cut shoulders or slots. Variants in the form of saddle and form mills may be used to cuts lengths of T-nuts or other complex shapes.

Figure 7. *A range of different horizontal-milling cutters. Photo courtesy lathes.co.uk.*

99

Figure 8. *A 10DP (diametral pitch) involute cutter suitable for forming imperial gears.*

Involute cutters
When making gears, an involute cutter will be needed in order to achieve the correct gear-tooth profile. Its form will depend on the diameter of the gear and its number of teeth. Unfortunately, each different module (metric) or DP (imperial) will need a different involute cutter.

Slitting saws
It is normal for slitting saws to be made from thin high-speed steel and to have a different number of teeth for a given size, depending on the metal to be cut. These saws are ideal for cutting deep, narrow slots or for parting off. However, they will need to be supported with a suitable arbor or mandrel described overleaf.

Figure 9. *A slitting saw mounted on a mandrel that is sitting in a v-block to stop it rolling.*

Figure 10. *A simple boring bar with an HSS cutting bit located near its tip.*

Drills
Twist drills are widely available in metric and imperial sizes made from HSS. Some are carbide tipped. The smallest metric drills are available in steps of one tenth of a millimetre; larger ones in steps of 0.25mm. Small imperial ones come in number and letter sizes. Number sizes may vary by as little as 0.001" and range up to just over $^{7}/_{32}$"; letter sizes vary by 0.006" or more up to $^{13}/_{32}$".

Boring bars and heads
The best way of boring a hole in a piece of metal depends primarily on the size of the hole required. For any diameter greater than 13mm (½") a boring tool is more practical than a drill. A little tool can bore quite small diameter holes. There are several different types of bar.

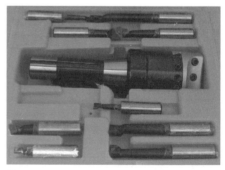

Figure 11. *A boring head with micrometer adjustment and interchangeable boring bars.*

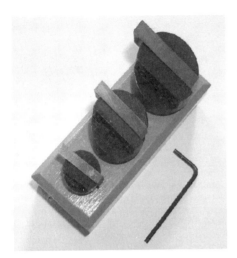

hones usually comprise three, sometimes two abrasive stones that can be rotated inside a cylindrical bore. External hones rotate stones around the periphery of a circular component. The hone is usually held in a chuck and rotated in a lathe but a milling machine may also be suitable.

Knurling tools

Knurling tools come in various types of design. The best are those that pinch both sides of the work piece rather than push against one side. This minimises the loads on both a lathe cross slide and the headstock. There is also a choice of knurling wheels that allow for diamond, diagonal or straight patterns to be cut in the metal work piece.

Figure 12. *A set of three different-size fly cutters and adjusting Allen key.*

The simplest involves a cutting piece held in the end of a length of metal like the one shown in Figure 10. A grub screw locks the cutter in place. Larger boring bars, with a micrometer adjustment, can alter the diameter of the hole to be produced. Boring bars, which can be used for working between centres, also normally use a cutting piece held in a long bar mounted between the head and tail stocks of a lathe.

Fly cutters

A fly cutter is the ideal tool for providing a top-class finish to a relatively large area of material. Its circle of sweep out to the cutting edge is normally adjustable and depends on the size of the tool and its radius. The cutting edge of the tool can easily be re-sharpened in exactly the same way as any lathe tool. Allen-key grub screws normally hold the cutter in place.

Hones

A circular hone is the right tool for getting a fine finish on circular objects. Internal

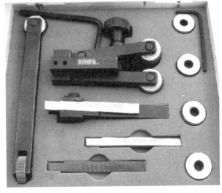

Figure 14. *A push knurling tool, left, and a pinch one, centre, with two pairs of spare wheels.*

Figure 15. *A cast-iron angle plate suitable for use on a lathe or milling machine.*

Holding ancillaries

Angle plates
The ability to hold work at 90° to a face plate, cross-slide or milling table increases

Figure 16. *A pair of arbors, one with a Morse taper, the other with a parallel shank.*

Figure 17. *An MT3 to MT2 sleeve. Photo courtesy Chronos.*

the flexibility of any machine tool. What is important when selecting an angle plate is its size, the provision and spacing of suitable slots or holes for attaching it (and items to it) as well as the accuracy of the angle and the flatness of the two faces. The dimensions needed depend primarily on the size of the lathe or mill on which it will be fitted.

Arbors/mandrels
Arbors or mandrels will hold work pieces or tools that are either to be rotated in a chuck or held in a tailstock. Some classic applications are to mount a Jacob's chuck, to clamp a slitting saw being rotated in a lathe or mill and to hold an item, which has first been bored, to allow for further machining. Arbors and mandrels can be obtained in various shapes and sizes, some of which are expanding, with either parallel or Morse-taper shanks.

Adapters, reducers and sleeves
On occasions, there may be a requirement to fit, for example, an MT2 arbor into an MT3 tailstock. A suitable adapter will be then needed. An alternative solution may be to use an appropriate arbor but with a parallel shank allowing it to be held in a chuck.

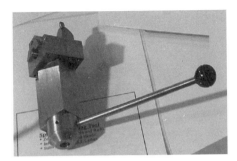

Figure 18. *The Hemingway ball-cutting attachment comes as a kit of parts with drawings and notes.*

Figure 19. *A 200mm (8") three-jaw scrolling chuck with a speed limit of 2,000rpm.*

Ball-turning attachments

When trying to turn a curved shape or the ultimate curve, a ball, a special type of attachment is normally needed. Every ready-made design involves the ability to rotate a cutting tool around a pivot fixed to the lathe. Figure 18 illustrates a typical solution that mounts on a cross slide and readily enables balls to be turned.

Chucks

There are many different shapes, sizes and types of chuck. The standard one that is fitted to most lathes is a 3-jaw self-centring chuck. Its diameter depends on the size of the lathe. There are other alternatives that will add to the capability of any lathe and the size of any chuck will impact on the size of component it can hold in its jaws.

The maximum size of chuck that a lathe can accommodate depends on its centre height or swing and also on whether it has a gap bed. And as well as being frustrated by a chuck that is too small to hold a large work piece, too large a chuck may not accurately hold a small component.

In addition to the conventional 3-jaw chuck, most lathe manufacturers offer a variety of other chucks as optional extras. Many lathes of Asian origin include more

than one chuck with the basic lathe. Companies like Toolmex and Pratt Burnerd manufacture quality chucks, while there are also many inexpensive and often less precise chucks of Far East manufacture. While chucks are available with 2, 3, 4 and 6 jaws, model engineers normally fit only 3- and 4-jaw ones.

Lathe manufacturers use various ways of mounting chucks, which must accurately be attached to a back plate. This may be supplied or need to be made. It must match the headstock spindle. A screw-on method is commonly used (the Myford Super 7 lathes spindles have an M42.5 thread of 2mm pitch) but with this type of

Figure 20. *If a chuck is not purchased from the lathe supplier, it is important to make or buy a suitable back plate to fit the chuck to the lathe.*

103

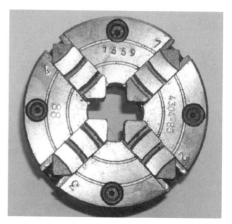

Figure 23. *An independent 4-jaw chuck.*

Figure 21. *A 3-jaw Camlock chuck with the cams fitted and chuck key in place. Photo courtesy RDG Tools Ltd.*

fitting the chuck can unscrew when the lathe is turning 'backwards'. Alternatives are to bolt the chuck onto a spindle plate or to employ self-locking devices such as the Camlock system. So the fitting may be:

a. Standard with a plain back.
b. Threaded with a specified thread size.
c. Type A1/A2 with a short spindle taper and one or two rings of bolt holes.
d. Type D1 for Camloc chucks.
e. Type L with a long-taper key fitting.
f. Type DIN with a short tapered flange to fit particular spindle sizes/thread.
g. British/ISO short taper that has a bayonet-ring fixing.

Figure 22. *A set of chuck outside jaws. The arrow shows jaw numbering.*

It is thus essential to establish which fitting applies before buying any chuck.

Chuck jaws are invariably numbered so that they can be put into the correct position on the chuck. A set of external jaws and soft jaws adds flexibility to a chuck.

Although not often an issue, every chuck has a maximum speed limit that must not be exceeded. If the lathe top speed is high, check the limit on any new chuck before making a purchase.

3 jaw

A scrolling 3-jaw chuck normally comes with sets of inside and outside jaws for holding different types of work piece. An optional set of soft jaws is useful when trying to hold malleable metals. The size of the centre hole in the chuck (and the lathe spindle) will limit the diameter of the work piece that can be passed through the jaws and may be a major issue when the time comes to turn an axle. A 3-jaw chuck can hold circular and hexagonal work but not square components. Furthermore, the accuracy with which it can hold a component centrally is often limited compared to an independent 4-jaw chuck.

Figure 24. *A collet chuck for a lathe or milling machine and a set of collets.*

Figure 25. *Left, a keyed Jacobs chuck with key, and right, a keyless one with an MT shank. Photo on right courtesy RDG Tools Ltd.*

Jacobs
A chuck that is often fitted to a lathe tailstock to hold drills and reamers is a Jacobs 3-jaw chuck. It is either opened and closed with a key that engages in a toothed ring around its periphery or by hand in the case of a keyless one.

Magnetic chucks
Most magnetic chucks are rectangular and designed for use when milling or grinding components, particularly thin ones. There are, however, a few round magnetic chucks for lathe use. The magnetic effect is normally created by permanent magnets. Turning a handle engages and unlocks the magnetic effect. However, in some chucks, the effect is generated by switching electro-magnets on or off.

4 jaw
For a square or any unevenly shaped component, a 4-jaw chuck is needed. Each jaw can move independently of the others, allowing awkwardly shaped parts to be securely held. The item being gripped should be carefully centred in the chuck using a dial gauge before turning commences. This type of chuck is also needed for offset turning, such as when making an eccentric for valve gear or a one-piece crankshaft.

4 jaw self-centring
A less common type of 4-jaw chuck is the self-centring variant that does exactly what its name implies. However, it will not hold a square item as accurately in the centre as a normal four-jaw chuck, but it is far quicker and easier to set up.

Collet
For the ultimate in centring accuracy, a collet chuck is required. However, such a chuck also needs a good range of collets to hold work of various diameters. This may involve buying a significant number of both metric- and imperial-size collets.

Figure 26. *A second-hand Eclipse 10" x 5" (254 x 127mm) magnetic chuck.*

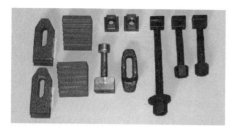

Figure 27. *A selection of clamping components for attaching work.*

Clamping sets

Any work piece mounted on a milling table, lathe face plate, top slide or vertical slide requires it to be securely held. Clamping sets contain many usefully shaped items. These include bolts or studs of different lengths with matching T-nuts to fit in a table's slots, spacers and nuts as well as slotted bars and step blocks to enable components to be held firmly in position.

Dividing heads

A dividing head is used to split a circle into a number of equal divisions. Some similar work can be done on a rotary table (see page 108). A dividing head consists of a spindle with a gear wheel and a shaft at right angles, with a worm gear and crank handle that allows the spindle to be rotated.

Figure 28. *A dividing head fitted with a chuck and installed on a milling machine.*

Figure 29. *A cast-iron faceplate with slots for attaching work. Photo courtesy Chronos.*

The spindle can carry a chuck to hold the work piece. An inter-changeable dividing plate and a spring-loaded plunger allow an indexing pin to locate into any one of a number of holes evenly spaced in the dividing plate mounted behind the crank. There is also a means of locking the spindle in position.

The crank has two sector arms that may be moved to show the right number of holes for any partial turn of the crank. This avoids any requirement to count holes when turning the crank.

The majority of dividing heads have a 40:1 reduction gearing, so forty crank turns revolve the spindle one complete turn. The indexing plate is a flat metal disk with a number of concentric circles of equally spaced holes. Several disks are likely to be needed to give a wide range of dividing options. The key factors in choosing a dividing head are:

1. Will it fit and can it easily be mounted on my particular lathe? It should also be readily installable on a milling table.
2. Will the range of dividing plates meet my gear-cutting or other dividing needs?

Face plates

For turning oddly shaped components, particularly castings, the use of a faceplate

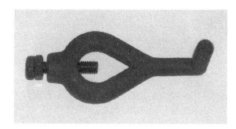

Figure 30. *A bent-leg dog transmits drive to work being turned between centres.*

Figure 31. *A 50mm (2") quick-release milling vice. Photo courtesy Chronos.*

is highly recommended. It is essential that that it will fit the lathe, both in terms of attachment and diameter. Faceplates come in many different sizes and with a range of slots or holes, or both, to mount work. They are often engraved with a number of concentric circles to help in the central location of the work piece. They will need a supply of correctly sized bolts or studs and nuts as well as clamping strips and short cylinders to raise the clamps as required. A typical faceplate is shown in Figure 29.

Lathe dogs
A lathe carrier or dog clamps around a circular work piece to transmit the lathe's rotary motion to the work. A dog is often employed when turning between centres. A straight-leg dog is used with a face plate that has a drive pin attached to it. The leg of the bent-leg type itself engages in the face plate. Of course the length of the lathe dog must match the faceplate and the lathe's centre height.

Machine vices (vises in the US)
A vice is one way of holding a component on a milling-machine table or the vertical slide of a lathe. Vices come in many forms; swivel, rotary head, rotary head with swivel, tilting, 3-way, universal, sine, quick release, low profile and self centring. Vices also come in many sizes that should

be a match to the dimensions of the lathe or milling machine as well as the size of work pieces. The best vices are made from close-grained high-tensile cast iron and have a precision-ground body with hardened sliding surfaces as well as an enclosed lead screw. Consideration should be given to the method used to attach the vice to the machine table, the height of the vice relative to the maximum distance between the milling head and the top of the vice and its ability to hold round work. Also, the action of the handle that

Figure 32. *A swivelling machine vice that can rotate through 360°.*

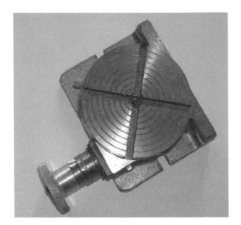

Figure 33. *Rotary tables can be mounted horizontally or vertically and used for a wide range of tasks.*

operates the moving jaw should be smooth and the jaw should always maintain its position parallel to the fixed jaw.

Rotary tables

While machines other than mills may make use of rotary tables, their function is particularly well suited to work being milled. The worm-wheel drive typically allows the table to be moved by as little as one-tenth of a degree and then locked

Figure 34. *The Arc Euro Trade 100mm (4") rotary table complete with stepper motor. Photo courtesy Arc Euro Trade.*

Figure 35. *A pair of screw jacks, one with the height-adjustment tommy bar.*

in position. Tables are available in a range of sizes, of which the most popular for model engineering range from 75mm (3") to 200mm (8"). Some rotary tables, such as the one shown in Figure 34, may either be fitted with or come complete with a stepper motor in order to provide a fourth CNC axis.

Screw jacks

Sometimes it is hard to hold a part in just the desired position. Small screw jacks may be used to ensure that clamping pieces do not rock. The key considerations

Figure 36. *A sine table set up on a milling machine table.*

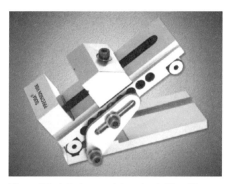

Figure 37. *A Soba precision sine table with built-in vice. Photo courtesy Chronos.*

are the minimum and maximum height of the jack, the area of its base and its ease of adjustment and locking.

Sine tables
A sine table can be used for accurate angle measurement as well as for holding work pieces at an angle for machining. Sine tables may be compound and/or magnetic and may include a built-in vice.

Steadies (rests in America)
When turning long thin pieces of metal, a steady is essential to stop the work piece flexing when it is being cut. There are two main types of steady that will achieve this.

A fixed steady attaches to the lathe bed itself and will support the work piece to prevent it from bending under the force applied by the cutting tool. However, it will provide support only at the fixed point where it is mounted.

A moving steady is bolted to the saddle and travels along the bed with the cutting tool. As both are attached to the saddle, the moving steady provides better support since it can be set always to be close to the point where the lathe tool is removing metal. This is also where the bending forces are potentially at their greatest.

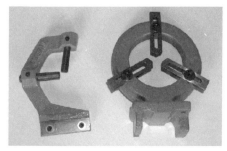

Figure 38. *Left, a moving steady, and right, a fixed one.*

Regardless of the type of steady, it must be compatible in terms of size and its method of attachment to the saddle or bed with the lathe on which it will be fitted.

Tailstock tooling
The tailstock of any lathe will hold many different accessories. Almost invariably they will be mounted on a Morse-taper mandrel, preferably No 2 or 3, though the smallest lathes may use No1. It is also important that any accessories match the tailstock Morse taper. Many lathes employ the same Morse taper in their headstocks allowing some of the accessories to be fitted there as well. There are different

Figure 39. *Fixed and moving steadies mounted on a lathe.*

Figure 40. *A modification to a lathe tail stock to provide lever operation.*

ways of moving any item in a tailstock towards or away from the headstock. The normal method is the use of a hand wheel. However, conversion to lever operation means that when drilling, clearing holes of swarf is a much faster task as the bit can rapidly be withdrawn.

Centres (centers in the US) for holding the work are very common. Both fixed and half centres are often carbide tipped to reduce wear and frictional-heating effects. However, the mandrel taper of all tailstock tooling must match that of the lathe itself.

Figure 41. *Three different types of centre are useful for lathe work; from left to right rotating, fixed and half.*

Figure 42. *The Hemingway tailstock die holder that can hold three sizes of die.*

Fixed centres
A low-cost solution is a one-piece fixed centre with integral Morse-taper mandrel that can be used to support the end of a work piece being turned in a lathe. But such a centre will require an adequate supply of lubrication to avoid overheating.

Half centres
A half centre has almost half of its pointed end removed to avoid clashing with the tool cutting the work piece. It is, however, a 'fixed' centre and will thus need plenty of lubrication.

Rotating centres
In a rotating centre, the part supporting the work is mounted in a housing with a suitable bearing, usually a ball bearing, to allow it to turn with the work. It is a much more elegant solution than a fixed centre but both types can limit access to the work piece at the 'centre' end.

Tailstock die holders
While a tap can be held in a Jacobs chuck, a die needs a special housing to hold it. Such holders are readily purchased but are not difficult to make in a lathe.

110

Figure 43. *An indexable tool holder can be a great time saver. This one is holding a live centre, a Jacob's chuck and a die holder. Photo courtesy Chronos.*

Indexable tool holders

An indexable tool holder comes into its own when making a number of identical parts that require several different tasks to be undertaken on each one, such as drilling and then tapping. The example shown in Figure 43 is mounted on a Morse taper for locating in the tailstock. It can house three different types of tool; in this case a live centre, a Jacobs chuck and a die holder.

Toolmakers' clamps

Often a work piece must be firmly fixed in place. Toolmakers' clamps are designed to prevent any rocking of the part being held during machining. While they are simple to make in the home workshop, there are many different types and sizes available from various specialist suppliers.

Tool posts

The simplest form of tool post holds just a single tool in position on the cross slide. But for quicker working a 4-way tool post will hold several tools at the same time

Figure 44. *A pair of small toolmakers' clamps.*

and may be rotated to change tools. This facility minimises the time taken to move between several different tools being used one after the other.

Another alternative is a quick-change, height-adjustable tool post where each tool is held in an identical fitting that can rapidly be attached to or removed from the tool post. Extra-long tool holders can be used to provide more clearance from the tool post.

A capstan or a quality four-way tool post will index into one of four preset positions; less expensive four-way tool posts will not provide exact repositioning of the tool but

Figure 45. *The classic Myford one-way tool post.*

Figure 48. *A capstan tool holder that mounts on a cross slide. Photo courtesy Home and Workshop Machinery.*

Figure 46. *A typical four-way tool post.*

the post can be locked in any position as it is turned allowing infinite adjustment of the tool position.

A rear tool post is often favoured for parting off. It holds the tool upside down. Thus, should the tool starts to dig into the work piece, its rotation tries to lift the tool, moving it away from the surface being cut and reducing its tendency to dig in.

It is important, particularly on smaller lathes, when choosing a new tool post, to be sure that it will allow the cutting edge of the various tools to be set on the lathe centre-line, rather than too high. This will apply whatever the type of tool post.

V-blocks

To securely hold round components, one or more V-blocks, complete with clamps that match, are helpful. The blocks come in a range of dimensions and different size V-grooves on opposite sides. A very high degree of manufactured accuracy is most desirable with all faces ground so they are truly flat and at 90° or at 45° to each other and opposite faces must be parallel. Precision-ground, matched and numbered in pairs should be exactly the same size to an accuracy of around 0.005mm (0.0002"). Both the blocks and their clamps should be strong enough to hold parts being machined.

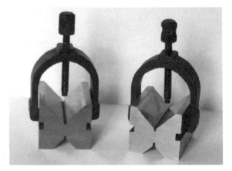

Figure 47. *A quick-change tool post and spare tool holders.*

Figure 49. *A pair of precision V-blocks and matching clamps.*

Figure 50. *A vertical slide for a Myford lathe. Photo courtesy Chronos.*

Vertical slides

Whether a lathe comes complete with a vertical slide or whether this is considered an accessory depends on the approach of the lathe manufacturer. The benefit of owning a vertical slide is that allows a work piece to be mounted on it so that milling, boring and drilling of non-round items can be undertaken in the lathe. The size of area that can be milled will depend on the amount of movement of the cross and vertical slides but will inevitably be rather more limited than on a milling machine.

Measuring equipment

To be able to obtain accurately machined components, some measuring equipment is essential.

Callipers

There are several types of callipers but those with digital readouts tend to be more popular than those with dials or verniers. The maximum distance that they can

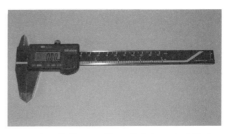

Figure 51. *Callipers with a digital readout that can be set to metric or imperial units.*

measure should be chosen to suit the size of models or full-size items that will be made. They can measure inside and out-side dimensions as well as the depth of holes. A typical digital calliper will have a resolution of 0.01mm (0.0005"), maximum total deviation of +/-0.04mm (0.002") and a repeatability of 0.01mm (0.0005").

Centre finders or wigglers

Different attachments, which are readily interchangeable in the wiggler body, adapt the tool to carry out more than just centre finding using the rod with a point. Each attachment is fixed by its ball end in the collet but with a loose grip that will permit adjustment of its angular position.

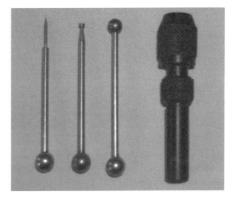

Figure 52. *A wriggler set with needle point, disc and ball contact probes, and the body.*

113

Figure 53. *A dial gauge on a magnetic base.*

Edge Finders

Work surfaces may be located easily and accurately with the edge finders; rods with a cylindrical or ball end. They will work with flat, straight edges, shoulders, grooves, round items and dowels.

Two different types depend either on mechanical contact or on an electronic sensor, which activates a light and sometimes also a buzzer, to indicate when an edge has been found.

Dial gauges or indicators

To ensure that a circular component is centrally located in a 4-jaw chuck, a dial indicator will show up any errors. It is one application when an analogue rather than a digital readout is preferable.

Figure 54. *A standard digital scale fitted to a lathe give a readout of saddle position.*

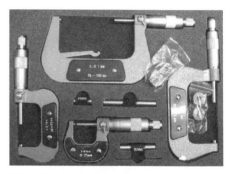

Figure 55. *A set of four analogue micrometers with verniers and complete with reference bars.*

Magnetic bases

To hold a dial gauge in position while making measurements, a magnetic base with a vertical rod and an adjustable arm is an ideal fixture. A rotatable knob (shown bottom right in Figure 54) allows the base to be de-magnetised enabling it to be moved to a new position and the base then re-magnetised to hold its position.

Digital readouts

Many lathes and milling machines come with digital speed readouts and can also be provided with displays of saddle and cross slide or table and spindle position. These are usually offered as optional extras but users may purchase separate units and fit them to their machine.

Micrometers

Accurate measurement to a resolution of 0.001mm (0.00005") requires the use of a micrometer, though for many less demanding measurements the accuracy of quality digital callipers will suffice. Conventional outside micrometers can measure from 0 - 25mm (0 - 1") while larger devices are readily available. Inside and depth micrometers are also useful for measuring internal diameters and the depth of holes. Micrometers can either

Figure 56. *A rule, three sizes of square and a centre square. Photo courtesy Chronos.*

Figure 57. *A Vertex coolant pump will need a reservoir and pipe work before being fitted to a lathe or milling machine that has a tray to catch the liquid for re-circulation.*

have an analogue readout using a vernier or a digital display. Different types of interchangeable anvils allow specialised measuring tasks to be carried out.

Rulers, squares and protractors

Perhaps the most basic measuring device is the steel rule, normally calibrated on one edge in millimetres and the other in inches. Various different sizes of square are also invaluable in setting up work pieces to ensure they are perpendicular to the surface to which they are attached. Additional items such as a centre square and a protractor for measuring angles are always useful.

Other items

Coolant systems

Some form of combined lubrication and liquid cooling is virtually essential if quality work is to be produced, particularly on harder metals. A system for pumping suds, a combination of soluble oil and water, is needed. This comprises a storage tank, a pump, pipe work, some of which must be adjustable to ensure the liquid is fed over the surface being cut, and a liquid-collection reservoir. Considerations

include the method of catching the suds after it has done its job, the type of pump to be used and the choice of pipe work.

A mains motor is the normal choice with a separate on/off switch and plastic coolant hose with a fully adjustable output nozzle, which is modular in construction and can easily be lengthened, shortened and pointed in a particular direction.

Figure 58. *A modular, flexible coolant hose with magnetic base that allows accurate fluid positioning. Photo courtesy RDG Tools.*

Figure 59. *A lamp on a flexible stem allows optimum positioning of the illumination.*

Figure 60. *A classic NVR switch that can be purchased as a stand-alone item.*

As an example, the Accura Vertex ACPS-009 coolant pump system may be fitted to a lathe or to a milling machine. It features a heavy-duty pump, a 9-litre (2-gallon) reservoir for coolant, a hose, a nozzle and a switch. The pump is mains powered and it consumes 94W ($^1/_8$hp).

Lights

Anyone working with a machine benefits from good illumination yet a lack of natural light is common in home workshops, especially on winter evenings. Many lathes and milling machines will include a repositionable low-voltage lamp; a very desirable choice for those without. Most machine-tool suppliers offer suitable lights as an optional extra and there are several specialist suppliers of purpose-made lamps. What is needed is the ability to point the light at the work; usually by a flexible arm between the base of the light and the bulb. A low-voltage light is best for safety. While a mains-powered light may be suitable, cables must be kept well clear of any machinery.

Lubricants

All machines require some form of lubrication and suitable oils and greases will be suggested and may be offered by the machine suppliers. Myford, for example, recommend Esso Nuto H32 (VG32) oil. Lubricants such as Multispec slideway oil are also useful. Specialist greases like those that are based on silicon, lithium and Teflon may also prove beneficial and model-engineering suppliers offer many fine lubricants; both oils and greases.

Soluble cutting oil is also required for any machine fitted with a coolant system although many model engineers prefer to apply a suitable oil by hand. Specialist cutting oils and compounds are widely available, such as Multicut, Neatcut, Rocol RTD and Trefolex.

NVR switches

Modern machines are invariably fitted with a 'no-voltage-release' (NVR) switch. For some older machines, the addition of an NVR in place of an existing manual switch is a good safety addition. This is because, in the case of a power cut without an NVR, if the machine was turning when the power failed, it will start to turn as soon as power is restored. NVRs are sold separately by some machine suppliers and also by a few ancillaries' dealers.

Conclusions

Hopefully, whether you are a beginner or an established model engineer, you will now be able to make an informed decision about which particular lathe, mill or multifunction machine to purchase. And if it is an area of interest, you may have found the right CNC solution. However, having made a significant investment in some workshop machinery, it is very important to protect your outlay.

Protection and care

Probably by far the greatest danger to any machinery housed in a home workshop is as a result of its environment. A large number of workshops are established in garages or wooden sheds where dampness and condensation can be major problems. Preventive maintenance must be carried out regularly on all machine tools. Both lathes and mills will usually be equipped with oilers and grease fittings so that the ways and other contact areas are kept continuously lubricated.

It is important to remember to apply regular lubrication to ferrous parts to avoid rust occurring. Background warmth and a cloth cover for each machine are also helpful solutions.

Smoke alarms & extinguishers

With valuable machines in a workshop, it is clearly sensible to install a smoke alarm (either battery or mains powered), to give early warning of any outbreak of fire. An equally important precaution is to provide a fire extinguisher in or near the entrance to the workshop. It should be a type that is suitable for dealing with an electrical fire. Extinguishers designed for use on vehicle fires provide a practical and relatively low-cost solution.

Insurance

With the value of any machine tool lying in the hundreds if not thousands of pounds or dollars, and with CNC equipment adding to the overall value, check that all the machinery is covered by normal household insurance, It may be necessary to have a separate policy to cover workshop items and to nominate individual machines if their value is high.

A policy that provides for replacement value is sensible. While it may be difficult for thieves to remove heavy items, lengthy absence from the home during holidays may provide just the opportunity.

Useful Contacts

Amadeal Ltd, Unit 20, The Sidings, Hainault Rd, Leytonstone, London, E11 1HD, UK. www.amadeal.co.uk

Arc Euro Trade, 10 Archdale St, Syston, Leicester, LE7 1NA, UK. www.arceurotrade.co.uk

Axminster Power Tool Centre Ltd, Unit 10, Weycroft Av, Axminster, Devon, EX13 5PH, UK. www.axminster.co.uk

(Birmingham Machines) All Machine Tools, 7537 S Rainbow Blvd, Suite 107-43, Las Vegas, Nevada 89139, USA. www.birminghammachines.com

Bolton Hardware, 3633 Pomona Blvd, Pomona, CA 91768, USA. www.boltonhardware.com

Boxford Ltd, Wheatley, Halifax, HX3 5AF, UK. www.boxford.co.uk

(Bridgeport) Hardinge Inc, One Hardinge Dr, Elmira, NY 14902-1507, USA. www.bpt.com

BriMarc Tools & Machinery, Unit 10, Weycroft Ave, Axminster, Devon, EX13 5PH, UK. www.brimarc.com

Ceriani Snc, Via Martini 629/651, 46030 Sustinente (Mantova), Italy. www.ceriani-mu.com

Chronos Engineering Supplies, Unit 14, Dukeminster Estate, Church St, Dunstable, LU5 4HU, UK. www.chronos.ltd.uk

Chester Machine Tools Ltd, Clwyd Cl, Hawarden Ind Pk, Hawarden, Nr Chester, Flintshire, CH5 3PZ, UK. www.chesteruk.net

Clarke International, Hemnall St, Epping, Essex, CM16 4LG, UK. www.clarkeinternational.com

(Colchester/Harrison) The 600 Group plc, Union St, Heckmondwike, West Yorkshire, WF16 0HL, UK. www.600group.com

Cowells Small Machine Tools, Tendring Rd, Little Bentley, Colchester, Essex, CO7 8SH, UK. www.cowells.com

CST, Scazzosi & C Snc, Via alle cave, 13 - 20029 Turbigo (Milano), Italy. www.cst-snc.it

EMCO Maier GmbH, Salzburger Str 80, A-5400 Hallein-Taxach, Austria. www.emcoworld.com

Excel Machine Tools, Colliery La, Exhall, Coventry, CV7 9NW, UK. www.excelmachinetools.co.uk

GOLmatic Machine Tools, Gottfried Prechtl, Auf der Aue 3, D - 69488 Birkenau, Germany. www.golmatic.de

GriffTek Technical Services, 7803 Puritan St, Downey, CA 90242, US. www.grifftek.com

Grizzly Industrial Inc, P.O. Box 2069, Bellingham, WA 98227, USA. www.grizzly.com

Harbor Freight Tools, 3491 Mission Oaks Blvd, Camarillo, CA 93012-6010, USA. www.harborfreight.com

Hemingway Kits, 126 Dunval Rd, Bridgnorth, Shropshire, WV16 4LZ, UK. www.hemingwaykits.com

HME Technology Ltd, Priory House, Saxon Park, Stoke Prior, Bromsgrove, B60 4AD, UK. www.hme-tech.com

Home and Workshop Machinery, 144 Maidstone Rd, Foots Cray, Sidcup, Kent, DA14 5HS, UK. www.homeandworkshop.co.uk

Jet Tools, Walter Meier (Manufacturing) Ltd, Bahnstrasse 24, 8603 Schwerzenbach, Switzerland. www.jettools.com

lathes.co.uk, Wardlow, Tideswell, Buxton, Derbyshire, SK17 8RP, UK. www.lathes.co.uk

LatheMaster Metalworking Tools, 2930 Belmont Av, Baton Rouge, LA 70808, USA. www.lathemaster.com

Machine Mart Ltd, 211 Lower Parliament St, Nottingham, NG1 1GN, UK. www.machinemart.co.uk

Microproto Systems, 12419 E Nightingale Ln, Chandler, Arizona 85286, USA. www.microproto.com

Model Engineers Digital Workshop, L.S.Caine Electronic Services, 25 Smallbrook Rd, Broadway, Worcestershire, WR12 7EP, UK. www.medw.co.uk

Myford Ltd, Wilmot La, Chilwell Rd, Beeston, Nottingham, NG9 1ER, UK. www.myford.com

Optimum Maschinen, Dr-Robert-Pfleger-Str 26, D-96103 Hallstadt, Germany. www.optimum-machines.com

Peatol Machine Tools, 19 Knightlow Rd, Harborne, Birmingham, B17 8PS, UK. www.peatol.com

Polly Model Engineering Ltd, Bridge Court, Bridge St, Long Eaton, Nottingham, NG10 4QQ, UK. www.pollymodelengineering.co.uk

Pratt Burnerd International, Park Works, Lister La, Halifax, West Yorkshire HX1 5JH, UK. www.pratt-burnerd.co.uk

PRO Machine Tools Ltd, 17 Station Rd Business Pk, Barnack, Stamford, Lincolnshire PE9 3DW, UK. www.emcomachinetools.co.uk

Proxxon-direct.com, Jaymac (Derby) Ltd, 852 London Rd, Derby, DE24 8WA, UK www.proxxon-direct.com

RDG Tools, Grosvenor Ho, Caldene Business Pk, Burnley Rd, Mytholmroyd, West Yorkshire, HX7 5QJ, UK. www.rdgtools.co.uk

Rejon Machine Tools, Mumby Lodge, Mumby's Drove, Three Holes, Wisbech, Cambridgeshire, PE14 9JT UK. www.rejon.co.uk

Rondean Machinery, Unit 20, Tanfield Lea Industrial Pk, Stanley, Co Durham, DH9 9QF, UK. www.rondean.co.uk

Sears, 3333 Beverly Rd, Hoffman Estates, IL 60192-3322, USA. www.sears.com

Sherline Products Inc, 3235 Executive Ridge, Vista, California, 92081-8527, USA. www.sherline.com

South Bend Lathe Co, PO Box 2027, Bellingham, WA 98227, USA. www.southbendlathe.com

Southern Tools, 4401 Northwest 37th Av, Miami, Florida 33142, USA. www.southern-tool.com

Taig Tools, 12419 E Nightingale La, Chandler, AZ 85286, USA. www.taigtools.com

Toolmex Corporation, 1075 Worcester St, Natick, MA 01760, USA. www.toolmex.com

Tormach LLC, 204 Moravian Valley Rd, Suite N, Waunakee, WI 53597, USA. www.tormach.com

(Wabeco) Walter Blombach GmbH, Am Blaffertsberg 13, D-42899 Remscheid, Germany. www.wabeco-remscheid.de

Warco, Warren Machine Tools, Warco House, Fisher La, Chiddingfold, Surrey, GU8 4TD, UK. www.warco.co.uk

WELsoft, 65 Daniells, Welwyn Garden City, Herts, AL7 1QT, UK. www.welsoft.co.uk

Index